T0177687

Morality and Mathematics

To what extent are the subjects of our thoughts and conversations real? This is the question of realism.

Many hold that, while there are real mathematical facts out there to be discovered, there are not real moral facts. Moral "facts," if there are any, are just the products of human invention. In this book, Justin Clarke-Doane argues that the situation is much more subtle and explores the similarities and differences between morality and mathematics, realistically construed. Although there are no real moral facts, morality is objective in a paradigmatic respect. Conversely, while there are real mathematical facts, mathematics fails to be objective. It follows from this that the concepts of realism and objectivity, which have been widely identified, are actually in tension. Our mathematical beliefs have no better claim to being self-evident or provable than our moral beliefs, and nor do our mathematical beliefs have better claim to being empirically justified than our moral beliefs. The book concludes with a general account of areas of philosophical interest. Clarke-Doane applies the realism/objectivity distinction across areas of inquiry and discusses its methodological upshot, broaching key topics of broad interest such as self-evidence and proof, the epistemological significance of disagreement, the philosophy/science comparison, metaphysical possibility, the fact/value dichotomy, and deflationary conceptions of philosophy.

"Justin Clarke-Doane identifies and explores the shocking parallelism between morality and mathematics: on a surprising number of philosophical fronts, the two disparate subjects seem to face common problems and analogous solutions and rebuttals, whether the issue is realism, a priori justification, objectivity, naturalism or pluralism. With consummate philosophical skill, Clarke-Doane teases apart the differences and uncovers what is truly common and what is not. A remarkable philosophical work."

Joel David Hamkins (University of Oxford, UK)

"Morality and mathematics would seem to be significantly divergent fields of inquiry. Justin Clarke-Doane is the rare philosopher with the requisite technical mastery of both fields to see past the superficial differences and highlight the important parallels lurking beneath, revealing how the issues of realism, objectivity, and justification we face in moral philosophy have close analogues in the foundations of mathematics, while noting what differences remain intact. This is a provocative and unique interdisciplinary contribution to how we understand truth and belief, with wide-ranging philosophical implications."

Sean Carroll (California Institute of Technology, US)

"Clarke-Doane's book offers a coherent and plausible set of answers to the notorious epistemological questions provoked by morality, and to the analogous questions that are provoked by mathematics. It is striking for its creativity, its rigorous arguments, its many subtle but important distinctions, its unusual breadth of expertise (covering philosophy of language, metaphysics, epistemology, philosophy of mathematics, and meta-ethics), and its rational control of a daunting battery of interactings considerations from these various branches of the subject. Exceptionally impressive philosophical talent and maturity are on display here. Needless to say, we probably haven't been given the final truth about these matters. But it's certain that anyone aiming to do better will have to grapple with Clarke-Doane's formidable arguments and conclusions."

Paul Horwich (New York University, US)

Morality and Mathematics

JUSTIN CLARKE-DOANE

OXFORD
UNIVERSITY PRESS

Great Clarendon Street, Oxford, OX2 6DP,
United Kingdom

Oxford University Press is a department of the University of Oxford.
It furthers the University's objective of excellence in research, scholarship,
and education by publishing worldwide. Oxford is a registered trade mark of
Oxford University Press in the UK and in certain other countries

First Edition published in 2020

Impression: 5

Published in the United States of America by Oxford University Press
198 Madison Avenue, New York, NY 10016, United States of America

British Library Cataloguing in Publication Data

Data available

Library of Congress Control Number: 2019953283

ISBN 978-0-19-882366-7

**Printed and bound in Great Britain by
Clays Ltd, Elcograf S.p.A.**

For Dad (1948–2017)

Contents

Acknowledgments

This manuscript has greatly benefited from the criticism of friends and colleagues. I received written comments on chapters 1-6 at a workshop on the manuscript hosted by Columbia University. Thanks to Wolfgang Mann for making that happen, and to Russ Shafer-Landau, Paul Boghossian, Mary Leng, Paul Horwich, Mark Balaguer, and Gideon Rosen for written comments on Chapters 1 to 6, respectively. Thanks to the excellent oral comments from participants, including, but not limited to, David Albert, Sean Carroll, Jessica Collins, Hartry Field, Joel David Hamkins, Tamar Lando, Jennifer McDonald, Michaela McSweeney, Thomas Nagel, Elliot Paul, Chris Peacocke, Lisa Warenski, and Crispin Wright. Thanks to Vera Flocke, David Kaspar, John Morrison, Olle Risberg, Katja Vogt, and an anonymous referee for meticulous comments on the entire document. Matthew Bedke, Colin Marshall, Conor Mayo-Wilson, and Michael Raven also offered penetrating commentary on a draft at an author-meets-critics symposium at the June 2019 Canadian Philosophical Association meeting. Thanks to Christopher Stephens for organizing that. Finally, I am grateful for the engaging discussion of chapters 3-6 which took place in Elizabeth Harman's and Sarah McGrath's Metaethics graduate seminar in Spring 2019.

Jens Haas, Brian Leiter, Peter Momtchiloff, Achille Varzi, and Katja Vogt guided me through the process. I deeply appreciate their encouragement. Thanks to Dan Baras, John Bengson, Betsy Clarke, Sinan Dogramaci, Emily Fletcher, Martha Gibson, David Keyt, Patricia Kitcher, Philip Kitcher, Karen Lewis, John Mackay, Farid Masrour, Christia Mercer, James Messina, Michael Ridge, Chris Scambler, Anat Schechtman, Mark Schroeder, Alan Sidelle, Alex Silk, Jon Simon, Neil Sinclair, Rob Streiffer, Bill Talbott, Mike Titelbaum, Jack Woods, and Justin Zacek for helpful discussion. Thanks to Matti Eklund for invigorating exchanges. Chapter 6 is indebted to them and his work. Thanks to Rani Rachavelpula for compiling the index. Thanks, finally, to Jenn McDonald, for everything.

Introduction

Philosophy is arguably the most inclusive discipline. Its subjects include everything from art to consciousness, from morality to mathematics. Moreover, it approaches its subjects from historical, conceptual, and scientific angles. Its breadth is not coincidental. Many take the defining aim of philosophy to be "to understand," as Wilfrid Sellars puts it, "how things in the broadest possible sense of the term hang together in the broadest possible sense of the term" [2007/1962, 369].

Despite its breadth, philosophy has become highly specialized. Corresponding to each of its myriad subjects are subdisciplines. Corresponding to morality is the subdiscipline of ethics, and corresponding to mathematics is the subdiscipline of the philosophy of mathematics. These subdisciplines, in turn, branch into sub-subdisciplines, including normative ethics, applied ethics, metaethics, and moral psychology in the one case, and mathematical logic, set theory, mathematical epistemology, and mathematical ontology in the other. To each sub-subdiscipline corresponds an enormous, often difficult, literature.

Specialization facilitates a kind of progress in philosophy. Problems are understood in depth, and theoretical options are developed with ever more sophistication. Such progress resembles Thomas Kuhn's normal science (Kuhn [1962]), minus the agreed-upon paradigm. But it also threatens progress toward the defining aim to which Sellars alludes. There is little hope of understanding "how things hang together" absent serious engagement between philosophy's diverse subfields—just as there is little hope of this absent serious engagement between philosophy and other fields.

0.1 Science and Value

Consider the question of *realism*. To what extent are the subjects of our thought and talk real? We all have a sense of it, prior to philosophical indoctrination. It is the question of whether the subjects of our thought and talk are "out there in the world" existing "independent of us." We are inclined to

Morality and Mathematics. Justin Clarke-Doane, Oxford University Press (2020). © Justin Clarke-Doane.
DOI: 10.1093/oso/9780198823667.001.0001

be realists about some areas while being anti-realists about others. For example, a common *naturalist* position among philosophers and scientists combines realism about the sciences with anti-realism about value. Naturalists, in the relevant sense, believe in independent facts about gene expression, plate tectonics, and quantum mechanics but do not believe in independent facts about what is morally good for us, what we epistemically ought to believe, or how prudentially we should live. Sean Carroll summarizes the naturalist's position on morality, in particular, as follows.

> There are not...moral truths...existing independently of human invention... but there are real human beings with complex sets of preferences. What we call "morality" is an outgrowth of the interplay of those preferences with the world around us, and in particular with other human beings. The project of moral philosophy is to make sense of our preferences, to try to make them logically consistent, to reconcile them with the preferences of others and the realities of our environments, and to discover how to fulfill them most efficiently.   [2010a, quotation marks removed]

Suppose we were to ask a naturalist why she takes different positions toward science and value. What would she say? Probably something like the following. First, facts about genes, the lithosphere, and electrons are implied by our best theories of the observable world, and those facts have been confirmed by observation and experiment. Second, knowledgeable individuals tend to agree on such facts, and, when there is a disagreement, there is a method— experiment and observation—by which to resolve it. Finally, we have at least a sketch of how human beings could acquire the knowledge of such facts that we take ourselves to have acquired. Facts about genes and so forth make causal marks on the world, marks to which our nervous systems respond.

By contrast, so-called "moral facts" would be different in all of these ways. Alleged facts about what is morally good or bad, right or wrong, obligatory or forbidden are not implied by any recognizably scientific theory. They are subject to endless controversy, even among people who agree on the non-moral facts, and who are otherwise intellectual peers. And there is no apparent method by which to resolve such disagreements. Finally, nobody has any idea how human beings could be reliable detectors of independent moral facts. Knowledge of such facts would be a mysterious extra kind of knowledge, over and above our knowledge of the natural world.

I have been speaking of the naturalist's attitude toward the empirical sciences, like physics and genetics. But a typical empirical scientific theory, rigorously formulated, presupposes pure mathematical facts as well. It presupposes, if only implicitly, whatever pure mathematical theories govern the mathematical entities to which it appeals. For example, Newton's law of universal gravitation presupposes real analysis, since the axioms of real analysis govern the numbers over which Newton's law quantifies. (A typical empirical theory is also closed under *logical consequence*—that is, if P is in the theory, and Q follows from P, then Q is in the theory—and has implications for how the world *would have been different* had initial conditions been different. In other words, a typical empirical theory presupposes logical and modal facts as well.) If a naturalist like Carroll were to declare that he is realist about, say, the standard model of particle physics, but *not* about mathematics, then it would not even be apparent what he meant.

So, an empirical scientific realist would seem to need to be a mathematical realist as well. She would seem to need to believe in independent facts about numbers, functions, and so forth, in addition to believing in such facts about genes, particles, and so on. As Hilary Putnam puts it,

> [Q]uantification over mathematical entities is indispensable for science…
> but this commits us to…the [independent] existence of the mathematical
> entities [that satisfy our theories]. This type of argument stems, of course,
> from Quine, who has for years stressed both the indispensability of quan-
> tification over mathematical entities and the intellectual dishonesty of
> denying the existence of what one daily presupposes.   [1971, 347]

However, unlike the contrast between independent empirical facts and independent moral facts, the contrast between independent *mathematical* (or, indeed, logical and modal) facts and independent moral facts is less straightforward. Even if mathematical facts are *implied* by well-confirmed scientific theories, it seems wrong to say that mathematical facts themselves have been confirmed. Was Riemannian geometry confirmed *as a pure mathematical theory* when general relativity was? That would seem to imply, falsely, that Euclidean geometry was disconfirmed when general relativity was. Similarly, while it can indeed appear that mathematics generates conver-gence, and that there is a method by which to resolve any remaining disagreements, this is questionable on inspection. Mathematical proofs pro-ceed from axioms. So, what they really show is that *if* the axioms are true, *then* so too is the theorem proved—at least assuming that there is agree-ment over the logic used. But moral claims admit of "proof" in this sense too. Gather together some claims from which the others follow and call them "axioms." What matters is how mathematical axioms compare to alleged "moral axioms." Is there disagreement over them? Do we have a method by which to resolve it? This is less clear. Finally, a longstanding objection to mathematical realism, famously pressed in Benacerraf [1973], is that it would be mysterious how humans could be reliable detectors of independent mathematical facts. We certainly do not interact with the likes of numbers and metric tensors!

Mathematics would, thus, appear to be a problem for the naturalist. Indeed, the outspoken naturalist, Alex Rosenberg, remarks, "[t]he criticism…that…I take seriously focuses on…our knowledge of mathematics—this is a serious problem for all naturalistic epistemologies" [2018]. On the one hand, it is not even apparent what it could mean to be a "realist" about our empirical scientific theories, while being an anti-realist about mathematics.[1] On the other hand, there may be no principled ground on which to be a realist about mathematics and an anti-realist about value. Whether naturalism, as that position is commonly understood, makes sense would thus appear to depend on whether one can be a mathematical realist and a moral anti-realist.

0.2 The Status of the Question

Can one be? The question has long interested philosophers. Plato (*Republic*, Book VII) closely associated mathematical knowledge with knowledge of the Good (Burnyeat [2000]), and the British rationalists belabored an analogy between simple mathematical and moral propositions (Clarke [2010/1705, 12]). Some philosophers have suggested that moral realism and mathematical realism "stand or fall together." Putnam begins a book with the declaration:

> [A]rguments for "antirealism" in ethics are virtually identical with arguments for antirealism in the philosophy of mathematics; yet philosophers who resist those arguments in the latter case often capitulate in the former.
> [2004, 1]

Putnam's remarks are characteristic of work on the issue. Despite their sweeping character, he does not defend them. The problem is specialization. Ethics and the philosophy of mathematics are such different subjects, and

---

[1] For more on this, see Section 3.5.

philosophy has become so specialized, that nobody really knows whether one can be a mathematical realist and a moral anti-realist. The "debate" over the relative standing of moral and mathematical realism has been mostly limited to trading impressions.[2] Most of them point in the opposite direction, as the following quotations illustrate.

A few philosophers claimed that we have a moral sense that perceives the moral rightness or wrongness of things....This theory might be worth taking seriously if morality were like mathematics. Mathematicians all agree that we know with certainty a large number of mathematical truths. Since experiment and observation could never be the source of such certainty, we...must have some other way of knowing mathematical truths—a mathematical sense that directly perceives them. For this argument to work in ethics, there would have to be little or no ethical disagreement to begin with. Since many moral disagreements seem intractable even among experts, the hypothesis that we are equipped to know moral truths directly is very difficult to sustain.   [Rosenberg 2015]

[M]athematics begins with a small number of shared, self-evident assumptions, while morality begins with a large number of inter-connected assumptions...all of which sound reasonable to the assumption-maker and precious few of which are truly self-evident. (In other words, moral epistemology is *coherentist* rather than *foundationalist*.)
[Greene 2013, 184–5, italics in original]

No, there is no such thing as a universal morality, and it is somewhat surprising that people are still asking this question in the 21st century. [I]f by "universal" we mean that morality is...like mathematical theorems, or perhaps like the laws of logic, then forget it.... [M]orality isn't even in the ballpark.   [Pigliucci 2018]

In explaining the observations that support a physical theory, scientists typically appeal to mathematical principles. On the other hand, one never seems to need to appeal in this way to moral principles. Since an observation is evidence for what best explains it...there is indirect observational evidence for mathematics. There does not seem to be observational evidence...for basic moral principles.   [Harman 1977, 9–10]

---

[2] There are exceptions, though no one has treated the matter in detail. See Brown [Forthcoming], Franklin [2014], Gill [2007], Kaspar [2015], Lear [1983], Lillehammer [2007], Parfit [2011] Scanlon [2014], and Wright [1994] for somewhat more sustained discussions of the comparison. See Leibowitz and Sinclair [2016] for a recent attempt to help rectify the situation.

In the case of mathematics, what is central is the contrast between practices or beliefs which develop because that is the way things are, and those that do not. The calculating rules developed as they did because [they] reflect mathematical truth. The functions of...morality, however, are to be understood in terms of well-being, and there seems no reason to think that had human nature involved, say, different motivations then different practices would not have emerged.   [Crisp 2006, 17]

Such one-off comparisons often betray serious misunderstandings. For instance, Peter Singer writes,

[Some moral realists] argued that there was a parallel in the way we know or could immediately grasp basic truths of mathematics.... This argument suffered a blow when it was shown that the self evidence of basic truths of mathematics could be explained in a different and more parsimonious way, by seeing mathematics as a system of tautologies, the basic elements of which are true by virtue of the meanings of the terms used. On this view, now widely, if not universally, accepted, no special intuition is required to establish that one plus one equals two -- this is a logical truth, true by virtue of the meanings given to the integers.... So the idea that intuition provides some substantive kind of knowledge of right and wrong lost its only analogue.   [1994, 8]

First, $1 + 1 = 2$ is not a logical truth (assuming that we mean first-order logic by "logic"). A countermodel is one in which the plus function maps 1 onto itself and to 3. Second, I am not aware of any contemporary advocate of the view that mathematics is a system of tautologies. Some logical positivists did suggest this. But their views were almost universally jettisoned after Kurt Godel proved the incompleteness theorems, and they were commonly ridiculed before that. Finally, far from being widely accepted, the notion of truth in virtue of meaning has been widely repudiated (Quine [1951b]). We may fix what proposition a sentence expresses. However, as Boghossian [1997 and 2003] emphasizes, we do not thereby fix whether the proposition expressed is true. Indeed, the idea of truth *in virtue of* meaning is dubiously coherent.[3]

_____

[3] Compare James Franklin, commenting on the same quotation from Singer: "That view is not universally accepted, nor widely accepted, nor indeed accepted at all by any living philosopher of mathematics" [2014, 198]. Singer appears to acknowledge some limitations of his remarks in a 2018 *AI Alignment Podcast* interview. For a contrary perspective on truth in virtue of meaning, see Russell [2011].

So, suggestions like the above are suggestive. But, as it stands, that is all they are. In order to see whether moral realism and mathematical realism stand or fall together, or whether ethics and the philosophy of mathematics have anything else to teach one another, we need to dig deeper. We need to bring ethics and the philosophy of mathematics into meaningful contact.

0.3 Overview of the Book

In this book, I explore arguments for and against moral realism and mathematical realism, how they interact, and what they can tell us about areas of philosophical interest more generally. I argue that our mathematical beliefs have no better claim to being self-evident or provable than our moral beliefs, contra the quotations from Rosenberg, Greene, and Pigliucci above. Nor do our mathematical beliefs have better claim to being empirically justified than our moral beliefs, contra the quotation from Harman. It is also incorrect that reflection on the "genealogy" of our moral beliefs establishes a lack of parity between the cases, contra the quotation from Crisp. In general, if one is a moral anti-realist on the basis of epistemological considerations, then one ought to be a mathematical anti-realist as well. And, yet, moral realism and mathematical realism do not stand or fall together, contra the quotation from Putnam. Moral questions—or the practical ones stake in moral debate—are *objective* in a sense that mathematical questions are not. But the sense in which they are objective can only be explained by assuming practical anti-realism. One upshot of the discussion is that the concepts of realism and objectivity, which are widely identified, are actually in tension.[4]

The book should be of interest to both ethicists and philosophers of mathematics. First, it shows that anyone who is a moral anti-realist on the basis of epistemological considerations ought to be a mathematical anti-realist too. Second, it raises problems for mathematical realism that have not been adequately explored. For example, it suggest that, in important respects, our mathematical beliefs are comparably contentious and contingent as our moral beliefs. Finally, the book reveals a special connection between the subjects of morality and mathematics. By comparing the subjects in detail, the correct philosophical account of each comes into focus.

---

[4] The sense of "objectivity" in question is similar to that of Field [1998a], and is opposed to relativism in the sense of Barton [2016] and Hare [1997]. See Section 1.6.

The book may also be of general interest. It concludes with a more encompassing account of areas of philosophical interest. There are those that are more like morality, such as normative epistemology and prudential reasoning, and those that are more like mathematics, such as modal metaphysics and (non-normative) logic. It is argued that, while we ought to be realists about the latter areas, they fail to be objective in just the sense that mathematics does. And while we ought to be anti-realists about the former areas, they are objective in the sense that mathematics is not. Along the way, key topics of general interest are broached, including: self-evidence and proof, the epistemological significance of disagreement, the philosophy/ science comparison, metaphysical possibility, the fact/value dichotomy, and deflationary conceptions of philosophy.

The structure of the book is as follows. In Chapter 1 I explicate (in Carnap's sense) the concept of realism, and distinguish it from related concepts with which it is often conflated. I show that, correctly conceived, realism has no ontological implications. One can be a realist without believing in any new entities. I also show that common objections to moral and mathematical realism fallaciously assume otherwise. One upshot of the discussion is that it is no response to Paul Benacerraf's epistemological challenge, mentioned above, to claim that there are no special mathematical entities with which to "get in touch." I conclude with a distinction between realism and objectivity, a distinction which is central to Chapter 6. Very roughly, objective questions are those which only admit of a single answer. By contrast, in a disagreement over a non-objective question, we can both be right. I use the Parallel Postulate, understood as a claim of pure geometry, as a paradigm of a claim that fails to be objective, even if mathematical realism is true. Conversely, I explain how realism about claims of a kind may be false even though they are objective in a sense that the Parallel Postulate is not.

In Chapters 2 and 3 I discuss how our mathematical and moral beliefs might be (defeasibly) justified, realistically construed, whether a priori or a posteriori. By "our mathematical and moral beliefs" I mean the range of mathematical and moral beliefs that we actually have, from trivialities of arithmetic to canonical theorems of set theory, from banalities such as "burning babies just for the fun of it is morally wrong" to egalitarian theses about gender and race. I depart here from much of the literature comparing morality and mathematics, both contemporary and historical, which has tended to focus on rudimentary claims of arithmetic and geometry.

In Chapter 2 I argue that our mathematical beliefs have no better claim to being a priori justified than our moral beliefs. In particular, they have no better claim to being self-evident, provable, plausible, "analytic," or even initially credible than our moral beliefs, despite widespread allegations to the contrary.[5] I consider the objection that pervasive and persistent moral disagreement betrays a lack of parity between the cases, and argue that there is no important sense in which there *is* more moral disagreement than mathematical disagreement, or in which moral disagreement is less tractable than mathematical disagreement. That is, there is no such sense which should lead us to conclude that our mathematical beliefs have better claim to being (defeasibly) a priori justified than our moral beliefs, realistically construed. A common argument to the contrary simply confuses logic—what is true if the axioms are—with mathematics (though I sketch a way in which one could also make a parity argument in the case of metalogic, the theory of what follows from what). I conclude with the suggestion that the extent of disagreement in an area, in any familiar sense, may be of little epistemological consequence—contrary to what is widely assumed.

Having argued that our mathematical beliefs have no better claim to being a priori justified than our moral beliefs, in Chapter 3 I argue that they also have no better claim to being a posteriori—that is, empirically—justified than our moral beliefs. I focus on Harman's influential argument to the contrary. Harman argues that since the contents of our mathematical beliefs are implied by our best empirical scientific theories, while the contents of our moral beliefs are not, only the former are empirically justified. I show that, on the contrary, Harman's reasons to think that the contents of our moral beliefs fail to be implied by our best empirical scientific theories serve equally to show that the contents of our mathematical beliefs do too, realistically construed. I then formulate a better argument for a lack of parity between the cases, in terms of indispensability. I argue that while the "necessity" of mathematics is no bar to developing a mathematics-free alternative to empirical science, contra an objection of Timothy Williamson, the contents of our arithmetic beliefs, realistically and even objectively construed, do seem to be indispensable to metalogic—the theory of what follows from what. But this would still only show that a subset of our mathematical beliefs have better claim to being empirically justified than any of our moral beliefs. And I argue that it does not even show that. Surprisingly, however, the range of moral beliefs that we have may be empirically justified, albeit in

[5] The relevant kind of analyticity is sometimes called "epistemic analyticity" and must be distinguished from the idea of truth in virtue of meaning mentioned by Singer in the quotation in Section 0.2.

a different way. Unlike mathematics, there may be no ground on which to rule out so-called "moral perceptions" as being on an epistemological par with ordinary perceptions ascribing high-level descriptive properties. I conclude with the prospect that there may be no principled distinction between intuition and perception, and, hence, between a priori and a posteriori justification.

Having shown that our mathematical beliefs have no better claim to being (defeasibly) justified than our moral beliefs, in Chapter 4 I consider attempts to undermine the latter by appeal to their genealogy—that is, Genealogical Debunking Arguments. I argue that, as standardly formulated, such arguments misunderstand the epistemological significance of explanatory indispensability. Debunkers observe that whether the proposition that P is implied by some explanation of our coming to believe that P is predictive of its having epistemically desirable qualities *when the fact that P would be causally efficacious if it obtained*. The problem is that these things are independent when the fact that P would be causally inert, and Genealogical Debunking Arguments assume otherwise. For example, when P would be causally inert, then whether the proposition that P is implied by some explanation of our coming to believe that P is independent of whether our belief that P is *safe* (that is, roughly, whether we could have easily had a false belief as to whether P), *sensitive* (that is, roughly, whether had it been that ~P, we would not still have believed that P), and (objectively) *probable*. I formulate a principle, which I call "Modal Security," which constitutes a criterion of adequacy for debunking arguments. It says that if such arguments are to undermine, rather than rebut, our targeted beliefs, they must give us reason to doubt their safety or sensitivity. But this is something that they do not do. Even if Modal Security is false, however, I argue that Genealogical Debunking Arguments have little force absent an account of the epistemically important quality that they are supposed to threaten. I conclude that the real problem to which Genealogical Debunking Arguments point is an application of the Benacerraf–Field challenge. The challenge is to explain the reliability of our moral beliefs, realistically construed. However, this challenge has nothing to do with whether the contents of our moral beliefs are implied by some explanation of our coming to have them.

In Chapter 5, I consider the Benacerraf–Field challenge, or what I call the "reliability challenge," in detail. After substantially clarifying the dialectic, I consider different ways of understanding the challenge. I begin with Benacerraf's preferred way, and then turn to improvements on it. I argue that none satisfies two key constraints which have been placed on the challenge. I then turn to more promising analyses, in terms of variations of the truths and variations of our beliefs. The best version of the former is the challenge to show that our beliefs are *sensitive,* in the above sense. This challenge is widely supposed to admit of an evolutionary answer in the mathematical case, but not in the moral. I argue that, on the contrary, the sensitivity challenge may admit of an evolutionary answer in the moral case, and not in the mathematical. But this is only because the sensitivity challenge is trivial to meet when the truths in question ascribe supervenient properties of concrete things, and impossible to meet when they do not. So this is an inadequate formulation of the challenge. This leaves analyses in terms of the variation of our beliefs. I argue that the best version of these is the challenge to show that our beliefs are *safe* in the aforementioned sense. Understanding the reliability challenge as the challenge to show that our beliefs are safe explains the otherwise mysterious conviction that, whatever its costs, the view that I will call "mathematical pluralism" at least affords an answer to the reliability challenge. Understanding the reliability challenge in this way also illuminates the epistemic significance of genealogy and disagreement. I conclude that whether the reliability challenge is equally pressing in the moral and mathematical cases depends on whether "realist pluralism"—or what I henceforth simply call *pluralism*—is equally viable in the two areas.

The rough idea to pluralism about an area, F, is that any F-like theory that we might have adopted is true of the entities which it is about, independent of human minds and languages.

In Chapter 6 I show that, while standard formulations of pluralism are dubiously intelligible, the view can be refined, and the resulting theory answers the reliability challenge for F-realism, *qua* the challenge to show that our F-beliefs are safe. It does so by giving up on the *objectivity* of the truths (in the sense of Chapter 1), but not on their mind-and-language independence. However, there is an essential difference between the mathematical and moral cases. Assuming mathematical pluralism, mathematical—as opposed to logical—questions get deflated. They become verbal in the sense in which the Parallel Postulate question is, understood as a question of pure mathematics. By contrast, assuming moral pluralism, all the pressing questions remain. If we call those questions *practical*, then we can frame the point as a radicalization of Moore's Open Question Argument. Practical questions remain open even when the facts, *including the evaluative facts*, come "cheaply." This means that mathematics and morality, insofar as it is practical, do differ, but the concept of realism alone is too crude a concept to do justice to the difference. Although practical *realism* is false, practical questions are *objective* in a paradigmatic respect. Conversely, while mathematical realism is true, mathematical questions fail to be. One upshot of the discussion is that the concept of objectivity, not realism, has methodological ramifications. Another is that the concepts of realism and objectivity (in one important sense of "objectivity"), which have been widely identified, do not only bifurcate. They are in tension.

I conclude by rehearsing key themes of the book and sketching their broader significance. I suggest a general partition of areas of philosophical interest into those which are more like mathematics and those which are more like morality. In the former category are questions of modality (counterfactual possibility), grounding, nature (essence), (non-normative) logic, and mereology. In the latter are questions of (normative) epistemology, political philosophy, aesthetics, and prudential reasoning. I argue that the former questions are like the question of whether the Parallel Postulate is true, *qua* a pure mathematical conjecture. They are verbal—but not because they are about words. They are verbal because reality is so rich as to witness any answer to them we might give. I illustrate this conclusion with questions of modality. I argue that, just as there are different concepts of geometrical point and line, all equally satisfied, there are different concepts of how the world could have been different. While it is, say, metaphysically impossible that you could have had different parents, it is logically possible that you could have, and there is nothing more "real" about metaphysical than logical possibility. In general, while typical questions of modal metaphysics are not about "possible," they might as well be. All we learn in answering them is how we happen to be using modal words, rather than learning what modal reality contains. By contrast, evaluative—or, more carefully, practical—questions are immune to deflation in this way. But the *reason* that they are is that they do not answer to the facts. So, their objectivity is not compromised if the facts are abundant. I conclude that the objective questions in the neighborhood of questions of modality, grounding, nature, and so on are practical questions as well. Practical philosophy should, therefore, take center stage.

1

Realism, Ontology, and Objectivity

This book is about arguments for and against moral realism and mathematical realism, how they interact, and what they can tell us about areas of philosophical interest more generally. But before I turn to those arguments, I need to say what "realism" about an area, in the pertinent sense, is supposed to mean. Of course, "realism" is a technical term, and we can define it how we like. But certain theses have been central to the debate over moral and mathematical realism. In this chapter I articulate a core notion of realism about an area, F, and explain its application to morality and mathematics. I then discuss several important theses that are independent of, though often conflated with, realism, one of which will be central to Chapter 6.

What follows is neither a conceptual analysis of the term of art "realism," nor an arbitrary stipulation for how to use the word. It is closer to an explication in the sense of Carnap [1950b, 3]. My aim is to locate a reasonably precise concept in the neighborhood of those that have been invoked in metaethics and the philosophy of mathematics which can serve as a useful point of departure for comparisons between the two areas.

1.1 Individuating Areas

Intuitively, if F is an area of inquiry, such as morality or mathematics, then F-realism is the view that typical F-sentences are true or false, independent of us, and that some substantive ones are true, interpreted at face value. So, mathematical realism is the view that some such sentences as "2 is prime" or "there are inaccessible cardinals" are true, and say what they seem to say, and similarly for moral realism. But this is very rough. What does F-realism come to, more exactly?

A preliminary question is: how are we individuating areas of inquiry? In the cases of concern, we can think of areas as individuated by their predicates—"is good," "is bad," "is a reason to," and so on in the case of morality, and "$\in$," "$<$," "is a number," and so forth in the case of mathematics—where predicates, in

Morality and Mathematics. Justin Clarke-Doane, Oxford University Press (2020). © Justin Clarke-Doane.
DOI: 10.1093/oso/9780198823667.001.0001

turn, are individuated semantically, not syntactically. (Many different theories of both sorts share this non-logical vocabulary, syntactically individuated. For instance, the language of prudence shares morality's vocabulary.) In the mathematical case, we can identify the area with all interpreted sentences containing *any of and only* the predicates in question. That will make the mathematical sciences, such as physics or economics, different areas from (pure) mathematics, as they should be. But in the moral case, we ought not rule out sentences which contain other predicates too. The important moral sentences are exactly those which link intuitively moral to non-moral predicates, such as "capital punishment is wrong." So, we should identify morality with all interpreted sentences containing *any* of the predicates in question.[1]

But exactly what predicates count as moral and mathematical? None of the arguments to follow turn on this difficult question. In fact, I will ultimately set aside moral predicates altogether, in favor of normative and evaluative ones. But it is worth flagging at the outset that the answer is likely indeterminate. Proceduralists will say that moral predicates correspond to norms that would be adopted by free, equal, rational, agents for mutual benefit, but functionalists may contend that those norms need merely regulate emotions like guilt and anger.[2] Moreover, historians will observe that our evaluative practices have transformed since antiquity (Frankena [1973, ch. 1]), and it is hard to believe that there was a magic moment at which they became "moral." It is tempting to at least require that moral predicates be *practical*. I will treat morality's practicality at length in Chapter 6. But it is unwise to build any particular thesis along these lines into moral predicates. Perhaps the standard way of characterizing morality's practicality—in terms of so-called *motivation internalism*—is highly controversial among moral realists.[3] For present purposes, we can probably do no better than to invoke Peter Railton's gloss. He writes,

> Various *substantive* concerns distinguish morality from other...domains.
> Moral thought has to do with acts and agents, with social practices and
> norms, and with matters of well-being and intrinsic value in a very general,

---

[1] If we wanted to be very careful, we would make sure to exclude sentences like "Jane believes that killing is wrong," as well as sentences like "Either killing is wrong or grass is green," from counting as moral sentences per se. We will have no need to be so careful, however.

[2] Thanks to Russ Shafer-Landau for discussion here.

[3] Motivation internalism says that an agent cannot sincerely judge that she ought to X without being at least defeasibly motivated to X. See Brink [1986].

encompassing sense.... Morality pays special attention to the conditions and principles needed to sustain reciprocal cooperation and mutual respect among agents in the pursuit of good lives.... [W]e also think that morality has a *practical* point. We expect ourselves and others to put moral judgments into practice and not merely to pay lip service to them. Moral judgment is thus associated with... "pro-attitudes" and positive motivation toward acts that fit with our moral judgments.

[2006, 203–4, italics in original]

The concept of a mathematical predicate is prima facie more determinate. While our conception of mathematics has also transformed since antiquity (Maddy [2008]), what it has transformed into—so-called "pure mathematics"—is a substantially different subject from that of the Greeks, or even Leibniz and Newton. Nevertheless, as physics becomes every more mathematized, there may turn out to be borderline physical/mathematical predicates too (see Section 3.5).

The above way of individuating areas of inquiry is not fully general. Consider modalities, such as metaphysical possibility. If we take the operator "it is possible that" as primitive, rather than regarding it as an abbreviation for the quantificational expression "there is a world in which," but also reject *de re* modal predications as unintelligible (as Quine [1953] did), then we cannot characterize metaphysical modality by its predicates. Indeed, logic has no characteristic predicates. Predicates like "is valid" or "follows from" are *meta*logical, not logical. A logical truth is a sentence in an ordinary language, such as "if snow is white, then snow is white." What ensures its status as a logical truth is that it is true under any interpretation of the non-logical vocabulary.

This problem will cause no confusion. There is no need for a once-and-for-all criterion of area-hood. If we are in doubt as to what sentences are in question, then we can specify them directly. Meanwhile, in the case of morality and mathematics, individuation by predicates suffices.

1.2 Bare-Bones Realism

Assuming that we know what we mean by an area of inquiry, F, we need to say what theses F-realism entails. I will state the theses with some redundancy in what follows (that is, the theses will not be independent), so as to highlight influential views which F-realism precludes.

Evidently, F-realism should at least entail that typical F-sentences are (determinately) true or false.[4] The "typical" qualifier allows that some F-sentences are neither true nor false thanks to vagueness or indeterminacy. But one is not an F-realist if one does not even believe that, barring vagueness and similar phenomena, F-sentences are apt for truth. Thus, F-realism at least entails:

[F-Aptness]   Typical F-sentences are true or false.

F-Aptness implies that austere forms of noncognitivism, such as A. J. Ayer's *emotivism*, according to which moral sentences are just used to express emotions, and (a common reading of) David Hilbert's *formalism*, according to which (nonfinitary) mathematical sentences are merely used to make moves in a game, are forms of anti-realism.[5] By contrast, it does *not* imply that sophisticated incarnations of noncognitivism which incorporate a deflationary theory of truth are too.[6] Following Railton [2006, 216, n. 6], I will call noncognitivist views *nonfactualist*.

How do we rule out sophisticated incarnations of nonfactualism from counting as realist? Blackburn [1990] suggests that, contra realism, sophisticated nonfactualists about morality hold that our (token) moral judgments are not explained with reference to their subject matter. But that is the *point* of at least one influential brand of "non-naturalist" moral realism (Dworkin [1996], Enoch [2011], Nagel [1986], Parfit [2011], Scanlon [2014])—the most uncompromising version of the view to which nonfactualism is supposedly opposed. Thomas Nagel writes,

> [I]t begs the question to assume that...explanatory necessity is the test of reality for values....To assume that only what has to be included in the best causal theory of the world is to assume that there are no irreducible normative truths.   [1986, 144]

It might be objected that if our moral beliefs were not "explained" by their subject matter—whether causally or in some other to-be-specified way—it would have to be a fluke that they were ever true (Street [2016]). In Chapter 4 I argue at length that this worry is confused. However, even if it

---

[4] I will mean determinate truth by "truth," for those who distinguish truth from determinate truth.

[5] See Ayer [1936, ch. VI] and Hilbert [1983/1926], respectively.

[6] See Horwich [1998] for a defense of deflationism about truth.

were sound, a non-naturalist can be a moral realist and a moral skeptic (see Section 1.4). So, it is more promising to appeal to the state of mind F-sentences are supposed to express. Unlike the sophisticated non-factualists, F-realists accept the following:

[F-Belief]   F-sentences conventionally express beliefs.

F-Belief rules out Blackburn's *quasi-realism*, according to which F-sentences express desire-like attitudes, and Gibbard's *expressivism*, according to which F-sentences express idealized plans.[7] (It does *not* rule out hybrid views, according to which F-sentences express both beliefs and noncognitive states (Fletcher and Ridge [2014]).) Of course, F-Belief only rules out such views if there is an informative account of the difference between beliefs and non-cognitive states. But, if there is not, then the schema is superfluous, because sophisticated nonfactualism does not get off the ground.[8]

In order to be an F-realist, one cannot just believe that typical F-sentences are true or false, and that they conventionally express beliefs. One must also believe that some are true. We should concede that sentences incorporating the predicate "contains phlogiston" are truth-apt. But, since we know that phlogiston does not exist, we ought not accept the truth of any sentence of the form "there is something that contains phlogiston and...." Actually, it is not enough to require that some F-sentence is true. We do not accept any sentence of the above form because we do accept its negation—we believe that it is true. What we are really after is the following:

[F-Truth]   Some atomic F-sentences are true.

In the case of morality and mathematics, F-Truth says that some F-sentences of the form "a is F," where "a" is a name, are true. So, F-Truth rules out John Mackie's *error theory* and Hartry Field's *fictionalism* from counting as realist views.[9] The former says that there are no morally good things, bad things, and so on, while the latter says that there are no numbers, functions, and so

---

[7] See Blackburn [1984 and 1993] and Gibbard [2003], respectively. See also Gibbard [1990]. See Schroeder [2010] for an overview and critique of the expressivist program.

[8] As Dreier writes, the more that sophisticated nonfactualists sound like realists, the less clear it is that they occupy a distinctive position. He writes, for example, "[a]ll the expressivists I know...would be happy to agree that there is no problem in ordinary conversation if we speak of "moral beliefs" and "ethical assertions" [2004, 28].

[9] See Mackie [1977] and Field [1980 and 1989], respectively. When the language in question lacks names, or first-order quantifiers (as with theories in the language of predicate functor logic), atomic sentences will look different.

forth. (Of course, both authors allow that it is *as if* there were the disputed things, in some or another respect.) Neither view precludes the truth of conditional claims such as *if* there are sets, *then* every set is well orderable (where "if…then" is the material conditional). On the contrary, if error theory and fictionalism are true, so that there are no morally good or bad things, or no mathematical entities, respectively, then every such conditional is (vacuously) true because it lacks a counterexample. Error theory and fictionalism preclude the truth of "substantive" sentences.

A paradigmatically anti-realist position about an area, F, combines F-Aptness, F-Belief, and F-Truth, with the view that the F-truths somehow depend on our minds or languages. Different versions of the position are generated by specifying different kinds of dependence. The F-truths could causally, counterfactually, or constitutively depend on our minds and languages. If there are forms of dependence not reducible to these, then there may be still other versions of the view. What counts as a relevant kind of dependence? Everyone should agree that the truth of many moral sentences counterfactually depends on human minds in the sense that, had there been no human minds, the sentence would not have been true. It would not have been true that Tom was wrong to lie, if Tom did not exist. So, this is an irrelevant kind of dependence. Presumably, all sides should also agree that the truth of "Mary has a good character" may constitutively depend on some of Mary's moral beliefs (not just on her existence). However, a realist will deny, while a constructivist may allow, that the truth of "Mary has a good character" constitutively depends on whether some person or group believes *that sentence* to be true. As Shafer-Landau puts it, the moral realist denies that our moral beliefs are stance-dependent. [10] So, we can add:

[F-Independence] The F-truths are independent of human minds and languages.

If F is morality, then F-Independence rules out Christine Korsgaard's *constructivism*, according to which (on one reading) the moral truths depend constitutively on what follows from a rational agent's "practical point of view."[11] Similarly, if F is mathematics, then F-Independence rules out L. E. J. Brouwer's *intuitionism*, according to which (on one reading) the mathematical truths depend counterfactually on what mental constructions we could perform.[12]

---

[10] See Shafer-Landau [2006, Pt I, § I].   [11] See her [1996].
[12] See his [1983/1949].

It is commonly assumed that F-Aptness, F-Belief, F-Truth, and F-Independence suffice to characterize a useful notion of F-realism. But this is incorrect. To see why, consider (one reading of) Gilbert Harman's *relativism*, according to which a typical moral sentence, "S," is really shorthand for the claim that, according to moral framework, M, S.[13] Then, given that whether *according to moral framework, M, S* is true, independent of human minds and languages, relativism may satisfy F-Aptness, F-Truth, F-Belief, and F-Independence. But relativism ought surely count as an anti-realist position.[14] According to it, there is no non-trivial fact as to whether M is true. Or consider Hilary Putnam's *if-thenism* (Putnam [1967]), according to which a typical sentence of, say, set theory, "S," is shorthand for the claim that, necessarily, any concrete system that makes the axioms of set theory true makes S true too. Then, again, given that such modal facts are independent of human minds and language, this view satisfies the above constraints. But it is anti-realist for much the same reason relativism is. It leaves no room for the question of what axioms are true, and allows that there are not really numbers, functions, and so on.

These considerations recommend supplementing F-Aptness, F-Belief, F-Truth, and F-Independence with:

[F-Face-Value]   F-sentences should be interpreted at face-value.

F-Face-Value is the semantic desideratum of Benacerraf [1973] generalized— that the truth-conditions of F-sentences should be taken to mirror their form. For example, the truth of "there are inaccessible cardinals" should depend on there being some objects, inaccessible cardinals—just as the truth of the sentence "there are tall buildings" so depends. The mirroring may not be exact. It is no part of realism—certainly not of moral or mathematical realism—that names are not disguised definite descriptions, à la Russell [1905], for instance. So, F-Face-Value ought to allow that truth-conditions of a sentence "a is H" are that there is exactly one F that is H.[15]

---

[13] See Thomson and Harman [1996]. Harman himself rejects this reading of his relativism in his [2012], but he is commonly (and I think naturally) read in this way. Harman's actual views are irrelevant to the illustration.

[14] On some readings, Harman is better interpreted as a pluralist (to be defined in Section 1.6), in which case he would indeed count as a realist on my taxonomy. See Vogt [2016, § 2.2] and Rovane [2013] for other interesting conceptions of relativism.

[15] The notion of a face-value interpretation is far from transparent, despite its ubiquity. Among other things, we must distinguish surface form from syntactic form. Some linguists would allow that "a is F" has Russellian syntactic form. It merely lacks this surface form.

Although it might be thought that the conjunction of F-Aptness, F-Belief, F-Truth, F-Independence, and F-Face-Value is all that one could ask for from F-realism, in fact it only generates a bare-bones version of the view. One cannot be an F-realist, in this sense, without believing that there are Fs, and that F-truths are independent of human minds and languages. So, one cannot be a mathematical realist without believing that there are mathematical entities and that mathematical truths are independent of human minds and languages, and one cannot be a moral realist without believing that there are "moral entities" (more on this below) and that moral truths are independent human minds and languages. But several key theses which are often identified with realism are, and should be, independent of it.[16]

1.3 Faithfulness

First, F-realism is consistent with *virtually any account* of F-objects and F-properties. I do not just mean that, for example, moral realism is neutral on the thorny question of whether moral properties count as natural. I mean that it is even neutral on the question of whether the property of moral goodness just is the property meditating in isolation. The "virtually" is thanks to the fact that we individuated predicates semantically, not syntactically, and, again, there are presumably *some* semantic constraints on what counts as a predicate of a given sort. If we had individuated predicates syntactically, then F-realism would be consistent with *literally* any account of F-objects or properties. Every consistent theory (in a first-order language) is true under some face-value interpretation, in a domain of pure sets, by the Completeness Theorem.[17] Had we individuated predicates syntactically, we would have had to concede that a Pythagorean, who identifies the subject matter of morality with numbers and moral properties with number-theoretic relations, may count as a moral realist.

---

[16] Although I am not discussing areas which resist individuation by predicates, let me note that it is unclear whether a face-value criterion is appropriate to the domain of modality. Arguably, Lewis [1986] is a modal realist (his view is, of course, called "modal realism," but this is not a pertinent use of "realism"). However, he denies that "<>P" is true at face-value, since he takes it to be a disguised quantificational claim. As is often said, there is no modality "in reality" for Lewis. Another case to consider is realism about colors. Self-described realists about colors would seem to deny Color-Independence, because they say that what it is for an object to be red is, roughly, for it to cause a certain kind of experience in observers. Despite this, such self-described realists hold that there "really are" colors Cohen [2009, ch. 1]. The problem is pressing in connection with mental entities generally. See Rosen [1994].

[17] The Completeness Theorem says that every first-order consistent theory has a (set) model.

Surely, it will be objected, this is strange. Why not add the following condition to F-realism?

[F-Faithfulness]   F-sentences should be interpreted faithfully.

The problem with F-Faithfulness is that there seems to be no way to characterize a faithful interpretation without prejudging what should be substantive questions. To illustrate, consider *reductionism* in mathematics. The mathematical reductionist holds that all mathematical entities are really entities of some one kind—usually pure sets.[18] The resulting interpretation of mathematical language satisfies Mathematical-Face-Value, as well as the other conditions. Is it faithful? It is certainly *surprising*. As Benacerraf [1965] argued, it is hard to even get one's head around the idea that the number 2 could *really be* $\{\varnothing, \{\varnothing\}\}$, as opposed to, say, the second Zermelo cardinal, or Frege-Russell class of all classes equinumerous with the one containing my hands. Indeed, Benacerraf took the idea that $2 = \{\varnothing, \{\varnothing\}\}$ to be sufficiently outlandish that he concluded that 2 could not be an object at all. For, he argued, if the number 2 were an object, then some reductive account of it would be true of it. But his argument was fallacious. Benacerraf overlooked the possibility that numbers could be irreducible entities. Indeed, this is just the conclusion that Moore [1903, Section 13] drew about moral properties on the basis of analogous considerations (Clarke-Doane [2008, 3, n.5]). And yet, even if we agree with Paul Benacerraf and G. E. Moore that the idea that there is a "hidden nature" to the number 2 and moral goodness is altogether too much to swallow (whether or not we take this to show that the number 2 and moral goodness could not be objects at all), mathematical and moral reductionists should not be classified as anti-realists. On one reading, Kurt Godel, the archetypical mathematical realist, was a mathematical reductionist, and W. V. O. Quine, the "reluctant Platonist," certainly was.

Indeed, *perhaps* we should not even rule out by stipulation that the property of moral goodness *is* identical to a number-theoretic property as anti-realist. Quine [1964] at least experimented with Pythagoreanism and Tegmark [2014] outright advocates it.[19] Maybe such theorists are better seen as extreme reductionists, rather than as fictionalists about everything but pure mathematics.

---

[18] See Paseau [2009] and Steinhart [2002] for contemporary defenses of set-theoretic reductionism. Bealer [1982] reduces them to properties.

[19] See Segal [2019] for a survey of contemporary Pythagoreanism positions.

The problem of faithfulness does not just arise in connection with reductionism. Consider fictional language, such as "Sherlock Holmes is 6 feet tall." How should we construe it? Anyone who knows the Holmes fiction will recognize that there is *some* important difference between "Sherlock Holmes is 6 feet tall" and "Sherlock Holmes is 4 feet tall." How should we think about that difference? One view, known as "fictional realism," says that the difference is that the former is true on a face-value interpretation, while the latter is false. Of course, not even fictional realists are so bold as to maintain that we might spot Holmes while strolling the streets of London. On the standard realist view, fictional entities are abstract, lacking spatiotemporal properties.[20] But if "Sherlock Holmes is 6 feet tall" is true on a face-value interpretation, in the sense of Fictional-Face-Value, then there is an object, Holmes, who satisfies the predicate "is 6 feet tall." Presumably no non-spatiotemporal object can be 6 feet tall (what would it mean to say that an object with no spatiotemporal properties has a certain spatial extension?). Hence, "is 6 feet tall," as it occurs there, must ascribe a property other than the one that is commonly taken to ascribe (such as the property of being *such that*, according to the Holmes fiction, Holmes is 6 feet tall). Is the resulting interpretation "faithful?" Meinongians would deny that it is (Reicher [2019]). They take the predicate to express the property it normally expresses. But they deny that *copula* expresses what it normally expresses. And standard realists would surely deny that *that* is faithful.[21] Evidently, some intuitively semantic appearance has to go. If "Sherlock Holmes is 6 feet tall" is true, under a face-value interpretation, in the sense of Fictional-Face-Value, then "is 6 feet tall" as it occurs in that sentence does not express the property that it standardly does. And if "is 6 feet tall" as it occurs there expresses the property that it standardly does, then "Sherlock Holmes is 6 feet tall" is not true under a face-value interpretation. The upshot is that if a faithful interpretation must honor *all semantic appearances*, then it would not even seem to be *possible* to be a fictional realist. The problem is actually quite general, and threatens the common wisdom that, whatever the epistemological or ontological costs of "Platonism"

---

[20] See, e.g., Wolterstorff [1979]. Similarly, Kripke [2011] argues that there are literally true predications of Holmes, though he denies that "Holmes is 6 feet tall" is among them.

[21] I assume throughout that, contra Meinongians, there is no such thing as a "lightweight existing." Indeed, the view that I call "pluralism" in Section 1.6 is an attempt to achieve some of the benefits of the view that there is without supposing that we know what it means to say that there are things that do not exist.

about an area, the Platonist can at least take the semantic appearances at face-value (Clarke-Doane [Manuscript]).[22]

The upshot is that the question of what kinds of things F-entities are is not settled, and is hardly even constrained, by an answer to the question of F-realism. This means that realists can disagree wildly about metaphysics. But this is as it should be.

1.4 Knowledge

Realism about an area, F, is also compatible with skepticism about the area—the view that we lack any atomic F-knowledge at all. Indeed, a familiar point is that the more realist one is about an area, F, the more serious the threat of F-skepticism. In Chapter 6 we will see that this truism is actually too simple. If the F-truths are not *objective*, in the sense that I will describe in Section 1.6, then the threat of F-skepticism may be mitigated, despite the truth of F-realism. For example, we will see in Sections 5.9, 6.1, and 6.2 that that threat is mitigated when F is the domain of pure mathematics. What is true is that, *if* the F-truths are objective, then F-realism and the assumption that we have atomic F-knowledge are in tension, for many choices of F— including both morality and mathematics. To the extent that the F-truths do not depend on our beliefs, there is, to that extent, the possibility that the truths fail to cooperate. However, this is, again, as it should be. Realism is a metaphysical doctrine. And metaphysics is one thing, and epistemology is another. So, F-realism is also neutral on the following commitment.

[F-Knowledge] We have atomic F-Knowledge.

But while F-realism does not entail F-Knowledge (or its negation), the combination of F-realism and F-skepticism is, of course, unappealing. This has been thought to be important by advocates of the epistemological challenges that we will consider in Chapters 4 and 5.

---

[22] This problem also infects Hilbert's formalism. Hilbert writes, "In number theory we have the numerical symbols 1, 11, 111, 1111 where each numerical symbols intuitively recognizable by the fact it contains only 1's....3 > 2...communicate[s] the fact that...the...symbol [2] is a proper part of the [symbol 3]." [1983/1926, 193] On pain of absurdity, Hilbert must have in mind symbol *types* rather than symbol tokens (Shapiro [2000, Section 6.3]). And types must be understood *Platonistically*, i.e., as existing independent of their instances, since otherwise it would be a doubtful empirical conjecture that every natural number has a successor. However, Platonic types are as opaque as numbers! They do not literally resemble their instances with respect to shape. They do not *have* shape. Nor is it evident, even if there are Platonic types, that 2 is a proper part of 3. For it is not evident that types have parts.

1.5 Ontology

Another feature of F-realism is that it need not imply that there are any new, or *peculiarly* F, entities—*even if it is non-reductionist*.[23] It might imply this. But whether it does depends on F. For example, mathematical realism evidently implies that there are peculiarly mathematical entities—numbers, sets, functions, and so on. They are peculiarly mathematical in that there is scant plausibility to the view that they are *also* ordinary physical, or otherwise uncontroversial, things. At best, they are properties (for example, cardinality or ordinality properties, and constructions out of them) exemplified by such things. By contrast, moral realism does not imply that there are peculiarly moral entities (though it does imply that there are moral entities), even given that moral properties cannot be reduced. While mathematics is about peculiarly mathematical entities, in the sense of naming or first-order quantifying over them, morality is about the likes of people, actions, and events. That there are such things is not in dispute in the context of the debate over moral realism. One might speak loosely and say that morality is about "moral properties." But, with rare exceptions, this is misleading speech. The sentence "Hitler is wicked" is not *about* wickedness, in the way that "2 is prime" is about the number 2. It does not refer to (or first-order quantify over) wickedness. It refers to (or first-order quantifies over) a person.

It is true that philosophers and scientists commonly frame the question of moral realism as the question of whether there are moral properties (Brink [1986, 24], Enoch [2011, ch. 3], Fitzpatrick [2008], Railton [1986, § I], Sayre-McCord [1988, § 3], Sturgeon [2006, 244]). Indeed, John Mackie says that "Plato's Forms give a dramatic picture of what objective values would have to be" [1977, 28], and Steven Pinker writes,

> [Moral truths] certainly aren't in the physical world like wavelength or mass. The only other option is that moral truths exist in some abstract Platonic realm, there for us to discover, perhaps in the same way that mathematical truths (according to most mathematicians) are there for us to discover.   [2008]

---

[23] I have already observed that realism about areas, in a pretheoretical sense of "areas," need not imply this, since, e.g., one can be a realist about metaphysical possibility but take modal operators as primitive (Melia [2014, ch. 4]). My present point is that even realism in the strict sense defined in this section need not imply this.

Presumably, no one really believes that mathematical truths per se are in a Platonic realm. It could be that *all* truths are in such a realm, simply because propositions or sentences are abstract objects. But Pinker clearly assumes that there is a difference between mathematics and, e.g., geography. His point is that mathematical *entities*, like numbers, are in a Platonic realm. The point of the comparison is to suggest that moral entities, like goodness, are too.

Such theorists might add the following schema:

[F-Ontology]   There are peculiarly F-entities (given that F-predicates cannot be reduced).

But this would be a mistake. It would imply that a moral realist could not be a Quinean nominalist about universals—that is, that she could not deny that there are literally properties over and above the concrete things that are propertied. It is no part of moral realism that, in order for a sentence "a is G" to be true, there must be an object, a, *and another object*, G-ness, and the former bears the exemplification nexus to the latter.[24] By that reasoning, a physicalist must be a realist about universals—even though the whole point of a prominent kind of physicalism is that there is nothing over and above the concrete things with which we causally interact.[25] A moral realist can hold that if "a is G" is true, then there is an object, a, and…it is G [Quine 1948].

What, though, about the rare exceptions alluded to above? There are cases where we seem to refer to (or first-order quantify over) moral properties. For example, "generosity is a virtue" is evidently about generosity, and this is a moral property. But, again, whether we should believe that *any* sentence of the form "F-ness is G" is true under a face-value interpretation is just the problem of universals. One could argue equally that there must be an object, The Red, over and above all the concrete red things, because "red is a color" is true. An obvious alternative is to hold that such sentences are useful shorthand for claims about concrete particulars, such as that, ceteris paribus, generous people are virtuous, or that red things are colored. (A similar point applies to apparent talk of "reasons," as in Parfit [2011] or

[24] I call it a nexus on account of a problem like that discussed in Plato [*Parmenides* 132a–b]. It cannot be a relation on pain of infinite regress—a is G only if a exemplifies G-ness, and a exemplifies G-ness only if the pair <a, G-ness> exemplifies exemplification....
[25] Field [1980 and 1989] appears to be such a physicalist.

Scanlon [2014].) Of course, an F-realist *need* not be a nominalist in the tradition of Quine [1948]. The point is that she can be one.

So, F-realism is not, on its own, an ontological thesis. It is an *ideological* thesis, in the sense of Quine [1951a]. It only becomes an ontological thesis in tandem with background commitments, such as that F-entities are not among the uncontroversial inhabitants of the concrete world, or that the truth-conditions of atomic sentences about that world appeal to properties. Although this qualification may seem pedantic, it forestalls two confusions.

First, it is commonly suggested that whether an epistemological problem arises for realism about an area, F, depends on whether it possesses a peculiar ontology. Indeed, canonical formulations of the Benacerraf–Field Challenge for mathematical realism (and Mackie's analogous "argument from epistemological queerness"), to be discussed in Chapter 5, suggest this (Benacerraf [1973], Field [1989, 26], Field [2005], Mackie [1977, 24]).[26] After all, if there are no peculiarly F-entities with which to commune, then what is the problem? But this thought, as natural as it may appear, is misguided. If it were right, then a moral realist could avoid the epistemological problem for moral realism by simply being a nominalist about universals, and a realist about metaphysical possibility could avoid an analogous problem by taking modal operators as primitive. On the contrary, we will see that, notwithstanding an influential tradition in the philosophy of mathematics and metaethics, it is no answer to the Benacerraf–Field challenge for realism about an area, F (or to Mackie's analogous argument), to respond, for instance, as Scanlon [2014, 122] does, that there are no F-entities with which to "get in touch."

Second, one might think that a self-identified F-realist who refuses to countenance peculiarly F-entities is somehow quietist. For example, Nagel [1997, Chapter 6] has been criticized for, on the one hand, advocating moral realism, while, on the other hand, denying that there are any peculiarly moral entities. Analogous criticisms have been leveled against Dworkin [1996], Parfit [2011], and Scanlon [2014]. But, given the above, there need be nothing quietist about this combination of positions. It could amount to a banal moral realism plus nominalism about universals.[27]

---

[26] Putnam [2012] also advertises his ontologically free modal interpretation of mathematics as a way to avoid Benacerraf's epistemological challenge.

[27] I take no stand on whether this *is* the position of Nagel [1997], Dworkin [1996], or Parfit [2011, Pt VI].

1.6 Objectivity

A final feature of F-realism, as defined above, is that it is neutral on what I will call the question of *objectivity*. The term "objectivity" is about as ambiguous as terms get. And I do *not* want to argue about how to use the word. But, as we will see, my choice of labels is not arbitrary.[28]

Consider a straight line. Now imagine that you pass two straight lines through that line, and on one side of the original line the sum of the angles that the two lines make with the original line is less than 180° – that is, it is less than the sum of two right angles. Must the two lines eventually intersect on the side which sums to less than 180°? Let us call the previous sentence the Parallel Postulate question. We could understand it as a question about lines in physical spacetime. In that case, the question is empirical, and the answer is either "yes" or "no." But let us ask the Parallel Postulate question as a question of pure mathematics—like that of whether there are infinitely many twin primes, or whether every set is well orderable. Is it true then?

The question is patently misconceived, even assuming mathematical realism. But why? The question need not be indeterminate. The terms "point," "line," and so on may have determinate meanings in a context. Nor need the question lack a mind-and-language independent answer. If mathematical realism is true, then how things are with the points, lines, and other geometrical objects depends entirely on the mind-and-language independent mathematical facts. The question is misconceived because, in a sense that comes apart from mind-and-language independence, it lacks a unique, or, as I will say, *objective* answer. There are simply different geometries (qua pure mathematical entities), and these give different answers to the Parallel Postulate question (syntactically individuated). In Euclidean geometry, the answer is "yes." But in, e.g., hyperbolic geometry the answer is "no." All we would learn in answering the Parallel Postulate question is something about us. We would just learn which geometrical structures we were talking about, as opposed to learning which such structure there were. Rather than asking "is the Parallel Postulate true?," we might as well just *stipulate* that we will use "point," "line," and so on to mean, for example, Euclidean point, line, and so on. Indeed, this is what we do.

Of course, paraconsistent logic to one side, no one can claim that a given use of the Parallel Postulate sentence is both true and false. Rather, different

---

[28] Indeed, the sense of "objectivity" in question is similar to that of Field [1998a], and is opposed to relativism in the sense of Barton [2016] and Hare [1997].

uses are true of different spaces. But it is not just that, say, Euclidean and hyperbolic geometries each have models—that is, that one can cook up a set-theoretic structure which satisfies them. Again, that is true of every (first-order) consistent sentence whatever. The geometries have *intended* models. They are satisfied by their intended subject. They are simply true. In this sense, geometry is unlike abstract algebra, and more like arithmetic. While the question of whether the axiom of commutativity for groups is true is also misconceived, this is because axioms of group theory are just about their class of models. Group theory with or without that axiom has no intended model.[29]

It might be thought that "objective" is out of place for the property I am discussing. "Objective" is often used to mean mind-and-language independent, or intersubjective, or having objects. That is not what I mean. There is a mind-and-language independent, intersubjective fact about a given use of the Parallel Postulate which obtains thanks to the arrangement of geometrical objects. I use "objective" because *while geometrical relativism (or if-thenism), as defined above, is false, it as if the most austere geometrical relativism were true*—assuming that mathematical realism is true. Where a relativist might claim that the Parallel Postulate is true according to the framework of Euclidean geometry, and false according to the framework of hyperbolic geometry, a realist will claim that that the Parallel Postulate (syntactically individuated) is true *of the lines*$_{Euclidean}$ and false *of the lines*$_{hyperbolic}$. Rather than switching frameworks on which to relativize, the realist switches intimately related subject matters. In both cases, the upshot is the same. "[T]he conflict between the divergent points of view...disappears...[B]efore us lies the boundless ocean of unlimited possibilities [Carnap 1937/2001, XV]."

It might be objected that it is not enough that various geometries are true, understood as pure mathematical theories. An eccentric metaphysician could always add that some one geometry, in addition to being true, is somehow privileged. Of course, no one would deny that we take interest in some geometries over others, or that some one geometry may be uniquely useful for a purpose (such as modeling physical spacetime). But there are

---

[29] Similarly, Joel David Hamkins writes, "At first...alternative geometries were presented merely as simulations within Euclidean geometry, as a kind of playful or temporary re-interpretation of the basic geometric concepts....In time, however, geometers gained experience in the alternative geometries, developing intuitions about what it is like to live in them, and gradually they accepted the alternatives as geometrically meaningful. Today, geometers have a deep understanding of the alternative geometries, which are regarded as fully real and geometrical." [2011, 11]

less pragmatic ways of understanding privilege. For instance, we could take "is privileged" to abbreviate "carves at the joints" in the sense of Sider [2011]. Ted Sider generalizes the notion of natural kind found in Lewis [1983] and applies it to everything from objects to quantifiers to operators.[30] If one pure geometry carves at the joints, then maybe the metaphysician could deny that Parallel Postulate question is misconceived. He could hold that it goes proxy for that of which geometry carves at the joints.

I will ultimately be arguing that this question, construed as factual, is no more objective than the Parallel Postulate question itself (Chapter 6 and Conclusion). Just as there are different concepts of point, line, and so on, there are different concepts of metaphysical privilege (Dasgupta [2018]). Even if, for example, Euclidean geometry is metaphysically privileged, it is not privileged*, for some privilege-like concept, privilege*. And while we could always ask which privilege-like property is *real privilege*, that is just the question of what "privilege" happens to mean out of our mouths. A perfectly analogous question arises in the geometric case. I will ultimately suggest that all manner of traditional philosophical questions, construed as factual questions, are analogous to the Parallel Postulate question in this way.

In fact, however, no one (as far as I am aware), whether realist or not, claims that some one (pure) geometry is metaphysically privileged. Even Platonists concede that different geometries have equal standing. Where theorists differ is over foundational theories, such as arithmetic and set theory. (By "foundational" theories I mean, roughly, theories in which one can carry out metatheoretic reasoning.) But the disagreement is not over whether a given concept of set, say, is metaphysically privileged. It is over whether it is satisfied at all. Intuitively, arithmetic is about *the* natural numbers, and set theory is about *the* universe of sets. There are not different natural numbers and different universes of sets (whether privileged or not) as there are different geometrical structures. Or so it is widely believed.[31] In effect, the question of whether the search for the true axioms

---

[30] David Lewis introduced his notion as a response to Nelson Goodman's New Riddle of Induction ([1983, ch. III, § 4]). David Hume pointed out that there is no logical entailment from the premise that every F has been G to the conclusion that the next F will be G. Goodman went further, noting that there is not even an *inductive support relation* between the premise that every F has been G, and the conclusion that the next F will be G, for many choices of G. According to Lewis, the support relation only holds when G is a natural kind. (Lewis also exploited his notion to resolve Putnam's Paradox of determinate reference, presented in Putnam [1980]. See Section 5.2.)

[31] For more on this, see Sections 2.2 and 6.2.

of mathematics is worthwhile is precisely the question of whether axioms of our foundational theories are relevantly like the Parallel Postulate (Hamkins [2015]). If *all* foundational axioms were like the Parallel Postulate, then it is hard to see what the practical difference would be between mathematical realism and the most radical form of if-thenist anti-realism, according to which a mathematical sentence, S, is just shorthand for the claim that, *if T, then S*—where T is a finite conjunction of a contextually specified set of axioms. Why not, then, also require that the F-realist believes the following?

[F-Objectivity] F-sentences are objective.

Again, not even the most objectivist-minded mathematical realist would accept this schema for all subareas of mathematics, like geometry. But some theorists would indeed hold that only one set theory (and arithmetic) has an intended (class) model. And, yet, it is unwise to build Set-Theoretic-Objectivity into mathematical realism. Even staunch objectivists, such as Kurt Godel and Hugh Woodin, take seriously the prospect that some set-theoretic sentences, like Cantor's Continuum Hypothesis (which says that there is a bijection between every uncountable subset of the real numbers and the whole set of them), are neither objectively true nor false.[32] They allow that there might be subtly different concepts of set, all independently satisfied. According to some of these, the Continuum Hypothesis is true, while, according to others, it is false. Moreover, no one such concept is metaphysically privileged—however exactly that might matter. Such a position could still be highly objectivist, entailing, in particular, that all sentences in the language of (first-order) arithmetic and analysis are objectively true or false. Moreover, it would still imply that there are sets and that truths about them are independent of human minds and languages. So, it would be exceedingly unorthodox to deem such theorists anti-realists. If Godel fails to count as a mathematical realist, then I am not sure that anyone counts as one.

We might try to weaken F-Objectivity, requiring only that some subset of F-sentences are objective. But there seems to be no principled way to do this. Between the extremes of "every sentence in the language of set theory is objectively true or false" and "every sentence in the language of set theory which is not a (first-order) logical truth or falsehood is true of its intended subject," there are myriad positions, each with defenders.[33] Moreover, not

---

[32] See Koellner [2014].
[33] See Gaifman [2012, Section 2] for an overview, although he is discussing the question of determinacy rather than objectivity. I distinguish these questions in Section 5.2.

even the latter position marks an absolute boundary in the anti-objectivist—or what I will call the *pluralist*—direction. It is formulated against the backdrop of classical logic. One could, in principle, go further and deny even that some classical logical truths, such that ~(G & ~G), where G is the Godel Sentence for Peano Arithmetic (PA), are objectively true (Priest [2012, § 5]).[34] How weak is weakest logic against which one could formulate an intelligible pluralist position? There seems to be no determinate answer—at least, if we are limited to logics that we can actually *use* as "all-purpose" logics (Clarke-Doane [2019, § 8]).

Although it may not be obvious, F-Objectivity has application to morality as well, even in the context of realism.[35] Consider, for instance, Richard Boyd's moral naturalism ([1988]).[36] According to it, our moral terms refer to whatever properties causally regulate their use. Such a view is perfectly realist. Indeed, it is a form of Cornell Realism (see Brink [1986 and 1989], Sturgeon [1984], and Railton [1986] for others). But it is also pluralist, since, as Horgan and Timmons [1992] emphasizes, it allows that there could be another community of people who use "good," "bad," and so on much as we do—to praise, blame, and evaluate conduct—though their use of these terms is causally regulated by different properties. If so, then while any given moral question may have a (determinate) mind-and-language independent answer out of our mouths, that answer may stand to the Parallel Postulate as the opposite answer stands to its negation. There would be a property—call it "goodness*"—which witnesses intuitively opposite verdicts on moral questions. I say "intuitively" since, of course, it strictly witnesses verdicts on moral* questions, just as hyperbolic space strictly witnesses the negation of a different postulate than the postulate which is witnessed by Euclidean space. Nor are such pluralist implications limited to naturalist views. Huemer [2005]'s moral intuitionism, and Scanlon [2014]'s neo-Carnapianism are similar in this way.

While such views raise difficult questions—questions that will be central to Chapter 6—they should not be ruled out as anti-realist. As Boyd himself notes, "[t]he subject matter of moral inquiry in each of the relevant communities would be theory-and-belief-independent in the sense relevant to the dispute between realists and social constructivists [1988, 226]." It will

---

[34] Objectively true and not also false, that is! Field [1998a] considers the prospect that G might be indeterminate.

[35] Eklund [2017] makes a related point, calling views which I call pluralist "alternative friendly." (He does not discuss the mathematical case.)

[36] The naturalist moral functionalism of Jackson [1998] would serve equally to make the point.

turn out to be a vexed question whether a moral realist can *be* an objectivist. It could be that the notions of realism and objectivity not only distinct, but are in tension. I will argue as much in Chapter 6.[37]

So, the question of F-realism should be distinguished from the question of F-Objectivity. Evidently, one can be an F-realist while denying that F-truths exhibit much objectivity. But there is no bar to holding that the F-truths exhibit a great deal of objectivity, even though F-realism is false. Consider, for instance, a mathematical anti-realist who accepts Mathematical-Aptness, Mathematical-Belief, Mathematical-Truth, and Mathematical-Independence, but takes a given mathematical sentence, "S," to be shorthand for the claim that *it is mathematically necessary that S*, where the operator is a logical primitive. Then, for all that has been said, it could be mathematically necessary that S just when "S" is true in the universe of sets, V, according to the objectivist realist. Such a combination of views would be objectivist, and anti-realist.

I have described a question that is not objective in the pertinent sense (the Parallel Postulate question), and have observed that certain accounts of an area, F, entail that typical F-questions are like it. But, in general, what does it mean to say that an area is, or is not, objective? It means that there are not, or there are, respectively, a plurality of F-like concepts, all satisfied, giving intuitively opposite verdicts to typical F-questions. ("Intuitively" because, again, divergent concepts give verdicts on strictly different questions.) More exactly, since objectivity can come in degrees, an area, F, is objective *to the extent* that there fail to be a plurality of F-like concepts, all satisfied. So, assuming Set-Theoretic-Truth, set theory fails to be objective to the extent that there are a plurality of set-like concepts that are satisfied. And morality fails to be objective to the extent that there are a plurality of goodness-like, obligation-like, reason-like, etc. concepts that are satisfied. Note that the question of objectivity, so understood, makes sense even when F-Belief fails—that is, even when *non-factualism* about F is true. It makes sense so long as the relevant notion of satisfaction (truth, reference, etc.) is deflationary, i.e., disquotational.

It is important to see that whether an area is objective is *independent of what the right semantics, or metasemantics, of it turns out to be*. Geometry is not objective because geometric—or geometric-like—reality is rich, if it

---

[37] Eklund [2017, 55] suggests that "The upshot [may be] not that ardent realism requires Alternative-unfriendliness. Instead, the upshot is that nothing could satisfy the ardent realist."

exists at all. It does not matter whether, in natural language, terms like "point," "line," and so on are ambiguous or indeterminate. Maybe we folk all mean Euclidean line by "line." If that showed that the Parallel Postulate was objective, then the fact that people generally mean simultaneous-relative-to-reference-frame-R by "simultaneous" would show that there is an objective fact as to what is simultaneous with what. But whether the theory of relativity is true does not turn on natural language semantics!

1.7 Conclusions

Realism about an area, F, as I will understand it, is the conjunction of F-Aptness, F-Belief, F-Truth, F-Independence, and F-Face-Value. F-Faithfulness, F-Knowledge, F-Ontology, and F-Objectivity are independent theses. The familiar Platonist position about mathematics combines mathematical realism with Set-Theoretic-Objectivity (and Mathematical-Ontology, which, again, follows from this in tandem with plausible assumptions). Similarly, Horgan and Timmons [1992] show that influential formulations of naturalist moral realism contravene Moral-Objectivity. In what follows, I will assume that the mathematical and moral realist is also an objectivist unless otherwise stated. But this assumption will require revisiting in Chapters 5 and 6.

Let me emphasize that the above characterization of moral and mathematical realism is not the only one conceivable. Some philosophers define realism about an area, F, in terms of such ideology as "fundamentality" or "grounding" (Dorr [2005], Fine [2001]).[38] While the main arguments to follow will in no way depend on my resisting such characterizations, I resist for two main reasons. First, many philosophers deny the intelligibility, or at least reality, of such relations as grounding. Acceptance of the notion should not be built into realism about another area. It would certainly seem that one could be a mathematical realist, say, without accepting the notion of ground. And while many theorists could be unwitting realists, that seems like an uncharitable interpretation of most of the theorists I will discuss in this book. Second, I will conclude with the suggestion that such domains as grounding are among those with respect to which we should be pluralists.

---

[38] Another interesting way of thinking about moral realism, in particular, is given in Marshall [2018].

There are a plurality of grounding-like notions, giving intuitively different answers to such questions as whether the grounding relation is well founded. But if grounding questions are not objective, then neither are questions of realism understood in terms of it. The suggestion that all manner of questions of traditional philosophical interest are relevantly like the Parallel Postulate question will be a central theme of the Conclusion.

2
Self-Evidence, Proof, and Disagreement

I have argued that realism about an area, F, is best understood as the conjunction of F-Aptness, F-Belief, F-Truth, F-Independence, and F-Face-Value.
One is a realist about an area, F, like morality or mathematics, when one
believes that sentences from the area conventionally express beliefs, and that
some atomic ones are true, interpreted at face value, independent of minds
and languages. This leaves open F-Faithfulness, F-Knowledge, F-Ontology,
and F-Objectivity. That is, it leaves open whether F-reality is anything like
we take it to be, whether we have atomic F-Knowledge, whether there are
peculiarly F-entities, and whether F-questions have objective answers—in
the sense that the Parallel Postulate question, understood as a question of
pure mathematics, does not (even according to the mathematical realist).

Nevertheless, comparisons between moral and mathematical realism
naturally take place against the backdrop of answers to some of these other
questions. In this chapter I begin with what is perhaps the most celebrated
context for the comparison, according to which, if there are atomic mathematical and moral truths at all, realistically construed, then they are objective, and our belief in them is justified a priori, that is, independent of
experience.[1] The question I consider is whether our moral beliefs have equal
claim to being so justified. I discuss a variety of problems with an affirma-
tive answer and argue that some are illusory, while others are of little consequence. I focus on the objection that mathematical propositions admit of
proof, while moral inquiry lacks a method, and the objection that there is
widespread and intractable moral disagreement but little or no such mathematical disagreement. I conclude with the suggestion that the extent of
disagreement in an area, in any ordinary sense of "extent," may have little
epistemological significance—contrary to what is widely assumed.

---

[1] I am referring to foundational mathematical truths, as opposed to truths of pure geometry
or abstract algebra. More on foundational theories in Section 2.2.

Morality and Mathematics. Justin Clarke-Doane, Oxford University Press (2020). © Justin Clarke-Doane.
DOI: 10.1093/oso/9780198823667.001.0001

2.1 A Priori Justification

Ask a philosopher for an example of an a priori belief, and she is likely to cite a mathematical belief. A priori beliefs are those that are justified—roughly, rational or reasonable—independent of experience.[2] (A *proposition* is said to be a priori if one could have an a priori belief in it.) In order to be justified in believing that, say, aspirin is protective against melanoma, one needs to gather empirical evidence. For instance, one might conduct, or consult, a longitudinal study. But in order to be justified in believing that, say, 2 + 2 = 4, nothing similar seems to be required. One can be justified in believing that 2 + 2 = 4, it is said, on the basis of "reflection alone."

Whether the notion of a priority marks a significant distinction is a question to which I return in Chapter 3. For now, what is important is whether the reasons cited for thinking that our mathematical beliefs are a priori justified serve equally to show that our moral beliefs are a priori justified—even if their counting as a priori justified per se does not matter in the end. According to an influential "rationalist" realist tradition, they do. Samuel Clarke writes,

> [T]is without dispute more fit and reasonable in itself, that I should pre-
> serve the Life of an innocent Man, that happens at any time to be in my
> Power; or deliver him from any imminent danger, tho' I have never made
> any promise to do so; than that I should suffer him to perish, or take away
> his Life, without any reason or provocation at all....For a Man endued
> with Reason, to deny the Truth of these Things; is the very same thing...as
> if a Man that understands Geometry or Arithmetick, should deny the
> most obvious and known Proportions of Lines or Numbers, and per-
> versely contend that the Whole is not equal to all its parts, or that a Square
> is not double to a triangle of equal base and height.   [2010/1705, 12]

More recently, Sarah McGrath writes:

> [W]e do not attempt to discover what people ought to do in particular
> circumstances by designing and performing crucial experiments; nor do
> we think that our moral beliefs are inductively confirmed by observation.

---

[2] Some authors distinguish the claims that belief in P is rational and the claim that belief in P is justified. But I will be using the claims interchangeably, since alleged distinctions between them will be irrelevant to my purposes.

Experience does not appear to play an *evidential role* in our moral knowledge. In these and other ways, moral knowledge seems to resemble mathematical knowledge more than it resembles the kind of knowledge that is delivered by the empirical sciences.   [2010, 108–9, emphasis in original][3]

McGrath's point is that we seem to arrive at some moral conclusions on the basis of reflection alone. Empirical evidence evidently bears on some moral questions. That Tom's action caused great suffering, and was performed with the intention to do this, bears on the question of whether that action was wrong—and whether Tom caused great suffering with this intention seems to be an empirical question. But whether it is wrong for a person to intentionally cause great suffering in the first place does not seem to be an empirical question in the same way.

2.2 Axioms and Proofs

So, some of our moral beliefs of the form "if x is F, then x is M," where 'F' is an intuitively descriptive predicate and 'M' is an intuitively moral one, seem to be a priori justified, if justified at all—just as our (pure) mathematical beliefs are widely supposed to be.[4] But they still might have less claim to being justified than our mathematical beliefs. After all, there is a striking difference between the cases. In the mathematical case, there appears to be an established method by which we can acquire (at least defeasibly) a priori justified beliefs. By contrast, there is no comparably palpable method for acquiring (even defeasibly) a priori justified moral beliefs. As James Rachels puts it, "[i]n mathematics there are proofs.... But moral facts are not accessible by...these familiar methods" [1998, 3].

In order to assess this contrast, we need to clarify what is meant by "proof." At first pass, a mathematical proof is a deduction of the theorem

---

[3] See also Peacocke [2004, 201].

[4] Note that while the mathematical truths which are plausibly a priori are just the pure ones (i.e., those containing only mathematical predicates), in the moral case "pure" claims (those containing only moral predicates) are typically devoid of any interest (see, again, Section 1.1). On the other hand, virtually everyone must concede that some impure moral claims are empirical. That is why empirical evidence—e.g., about what contributes to happiness—bears on moral theories in a way that it does not seem to bear on mathematical ones. So, the "moral rationalist" should claim that a proper subset of "impure" moral truths are a priori, and from these, plus empirical data, we arrive at a comprehensive moral theory. Indeed, this is exactly the position advocated in Peacocke [2004]. I will have more to say about the justificatory structure of apparently a priori theories, both mathematical and moral, below.

proved from some claims labeled "axioms." This is not quite right, however, because, to a philosopher at least, this will suggest that every statement in a mathematical proof is either an axiom or follows from previous lines by a rule of formal inference, like modus ponens. Mathematical proofs are not remotely so pedantic. Mathematical proofs are better thought of as *sketches*, or *abbreviations*, of deductions, in the philosopher's sense. They serve to convince the cognoscenti that there *is* a (Platonistic) proof in the philosopher's sense—or, less Platonistically, that there *could be* such a (concrete) proof.[5]

So far, the availability of proofs in mathematics does not distinguish it from any other subject. *For any claim whatever*, P, there is a "proof," in the philosopher's sense, of P from some "axioms." Just let the axioms be {P}! Evidently, if there is an epistemological disanalogy between our moral beliefs and our mathematical beliefs, it concerns the standing of the axioms.

What are the axioms in mathematics? There are different axioms for different areas of mathematics—including axioms for group theory, geometries of different sorts, analysis and arithmetic. In some of these cases, like group theory, the axioms do not even pretend to characterize a unique (up to isomorphism) intended model. There is no serious question as to whether the axiom of commutativity for groups is true, for instance. But in other cases, like analysis and arithmetic, the axioms do seem prima facie to answer to such a model. This is not just a baseless hunch. Kurt Gödel's Second Incompleteness Theorem implies that, if standard arithmetic, Peano Arithmetic (PA), is consistent, then so is PA conjoined with (a coding of) the claim that PA is *not* consistent, ~Con(PA). So, if arithmetic were like group theory, then the question of whether PA was consistent would be like that of whether the axiom of commutativity for groups is true! In fact, the same would be true of any theory that *interprets* PA, such as set theory—where, roughly, one theory, T, interprets another, T*, when we can translate T* into the language of T such that, if S* is a theorem of T*, then its translation, S, is a theorem of T, and the translation is reasonable. I do not just mean that PA, and theories interpreting it, might be consistent relative to one logic and inconsistent relative to a wacky alternative. I mean that there would be no objective question as to whether PA is *classically consistent*— that is, as to whether there is a proof of a contradiction in classical logic

---

[5] What is the modality here? Good question. This is one reason to think that some mathematics, realistically and even objectively construed, is indispensable to metalogic—or, at least, that nothing is gained by trading it for the primitive ideology of consistency. I return to this issue in Section 3.4.

from the axioms of PA.[6] Given that there *is* such a question (a matter to which I return in Sections 3.5 and 6.2), arithmetic and set theory exhibit some objectivity. Once this is granted, many go further. Once we concede that some arithmetic and set-theoretic questions have objective answers, maybe we should say that all do.

Despite being a relatively fringe area of pure mathematics, set theory is of special philosophical significance. Although it only has one non-logical predicate, "∈," the axioms of all other branches of mathematics (not just arithmetic) can be interpreted in it. In other words, those axioms can be understood as claims about sets in disguise. Mathematics could "get by" with an ontology of sets alone. As Herbert Enderton puts it, "mathematical objects (such as numbers and differentiable functions) can be defined to be certain sets. And the theorems of mathematics (such as the fundamental theorem of calculus) then can be viewed as statements about sets" [Enderton 1977, 10]. To be sure, as Benacerraf [1965] emphasizes, this observation does not show that, for example, the natural numbers *really are* sets. But it shows something important. It shows that *if the axioms of set theory are consistent*, then so too are the other axioms of mathematics.

Of course, the axioms of set theory, and of other branches of mathematics, such as arithmetic and analysis, could be consistent but false—at least in an epistemic sense of "could." If we had no reason to believe that those axioms were also true, then we would still lack a contrast with the moral case. We are rarely in doubt about the *(first-order) consistency* of a moral theory. What worries us is its truth. We might worry about the theory's Kantian "consistency." That is, we might worry whether we can will the theory to be universal law without somehow defeating the purpose of acting on it. But that is a much stronger property. All manner of Kantian "inconsistent" theories are (first-order) consistent. If a mathematical proof merely established the (first-order) *consistency* of the theorem proved, then, again, our moral beliefs would admit of analogous proof. We can establish the consistency of a moral theory in exactly the way that we establish the consistency of a mathematical one—by interpreting it in set theory.

The non-skeptical mathematical realist does not just believe that the axioms of set theory are consistent, however. She also believes that those axioms are true. Indeed, realists commonly argue that our epistemic *ground* for believing that the axioms of set theory are consistent is that they are

---

[6] The more exact analogy is to pure geometry, for reasons discussed in Section 1.6. See Section 6.2. I ignore this complication here.

true, and truth implies consistency (Frege [1980/1884, 106]). For example, Hugh Woodin writes of the proposed axiom, Projective Determinacy (which says that all projective sets are determined):

> For me, granting the truth of the axioms for Set Theory, the only conceivable argument against the truth of [Projective Determinacy], would be its inconsistency. I also claim that, at present, the only credible basis for the belief that the axiom is consistent is the belief that the axiom is true.
>
> [2004, 3]

But what is our ground for believing that the axioms of set theory are true? Since arithmetic and analysis can be interpreted in set theory, I will focus on the axioms of set theory.

2.3 Self-Evidence

An influential answer outside the philosophy of mathematics is that our ground for thinking that the axioms of set theory are true is that they are axioms in Euclid's sense—they are "intuitively obvious" or "self-evident." For example, Joshua Greene writes:

> Axioms are mathematical statements that are *self-evidently* true....If morality is like math, then the moral truths to which we appeal in our arguments must ultimately follow from moral axioms, from a manageable set of self-evident moral truths.   [2013, 184, italics in original]

Perhaps this is plausible in some cases. Even Kurt Godel once proclaimed that "the axioms force themselves upon us as being true" [1947, 483–4]. Consider the Axiom of Extensionality, $\forall x \forall y \forall z[(z \in x \leftrightarrow z \in y) \rightarrow (x = y)]$. This says that if "two" sets have the same members, then they are really one and the same (the converse is a logical truth in first-order logic with identity). Perhaps this is some kind of truism about sets.[7] Similarly, the Axiom of Pairing, $\forall x \forall y \exists z \forall w[w \in z \leftrightarrow (w = x \lor w = y)]$, says that, whenever you have "two" (perhaps not distinct) sets, there is another containing just those

---

[7] Although set theory without Extensionality has been explored and sometimes even advocated: (Scott [1961], Friedman [1973], Hamkins [2014]).

"two." This also seems hard to deny—though, even here, one might doubt that any existential statement, even conditional, could be self-evident.

We can imagine an analogous position in the moral case. Of course, there are no "moral axioms." But there are moral principles, and one could always regiment them (if one really had nothing better to do) and, from them, prove "moral theorems." The quotation from Clarke above suggests that he takes some to have equal claim to being self-evident. Indeed, he adds,

These things are so notoriously plain and self-evident, that nothing but the extremest stupidity of mind, corruption of manners, or perverseness of spirit, can possibly make any man entertain the least doubt concerning them.   [2010/1705, 12][8]

It is natural to respond to such suggestions with incredulity. Surely it is an abuse of language to deem any non-trivial moral proposition self-evident!

---

[8] Similarly, Thomas Reid writes, "From [these self-evident first] principles...the whole system of moral conduct follows so easily, and with so little aid of reasoning, that every man of common understanding, who wishes to know his duty, may know it. The path of duty is a plain path....Such it must be, since every man is bound to walk in it. There are some intricate cases in morals which admit of disputation; but these seldom occur in practice; and when they do, the learned disputant has no great advantage....In order to know what is right and wrong in human conduct, we need only listen to the dictates of our conscience, when the mind is calm and unruffled, or attend to the judgment that we form of others in like circumstances" [1983/1788, 640]. Locke made similar remarks. He writes, "[Moral propositions i]f duly considered and pursued, afford such foundations of our duty and rules of action as might place morality amongst the sciences capable of demonstration: wherein I doubt not but from self-evident propositions, by necessary consequences, as incontestable as those in mathematics, the measures of right and wrong might be made out, to any one that will apply himself with the same indifferency and attention to the one as he does to the other of these sciences. The relation of other modes may certainly be perceived, as well as those of number and extension: and I cannot see why they should not also be capable of demonstration, if due methods were thought on to examine or pursue their agreement or disagreement. 'Where there is no property there is no injustice,' is a proposition as certain as any demonstration in Euclid: for the idea of property being a right to anything, and the idea to which the name 'injustice' is given being the invasion or violation of that right, it is evident that these ideas, being thus established, and these names annexed to them, I can as certainly know this proposition to be true, as that a triangle has three angles equal to two right ones. Again: 'No government allows absolute liberty.' The idea of government being the establishment of society upon certain rules or laws which require conformity to them; and the idea of absolute liberty being for any one to do whatever he pleases; I am as capable of being certain of the truth of this proposition as of any in the mathematics" [*An Essay Concerning Human Understanding*, Book IV, ch. 3, § 18]. W. D. Ross also held that "the general principles of duty come to be self-evident to us just as mathematical axioms do" [1930, § II]. But he maintained that "there is an important difference between rightness and mathematical properties...Moral acts have different characteristics that tend to make them at the same time prima facie right and prima facie wrong" [ibid.]." The "prima facie" hedge is not needed in all moral propositions that have claim to being self-evident, however. I avoid it in the examples in Section 2.5. See Gill [2019] for further discussion.

Perhaps it is true that "cases in morals which admit of disputation" rarely affect our day-to-day affairs. We rely on those around us to at least *act in accord with* a wide array of moral norms as we go about our business. But, by itself, that does not show that there is widespread agreement about the *truth* of non-trivial moral norms.

In fact, however, the same is true of *non-trivial* mathematical principles. Even if such banalities as Extensionality and Pairing are self-evident, these do not even imply the existence of a single set! Set theory only gets going with the Axiom of Infinity, $\exists x(\varnothing \in x \,\&\, \forall y(y \in x \longrightarrow y \cup \{y\} \in x))$, which says that there is an inductive set—i.e., a set that includes 0, and includes the number, n+1, whenever it includes n. Even bracketing worries about the self-evidence of existential statements, it would be stretching the meaning of "self-evidence" to the point of uselessness to claim that it is self-evident that something *infinite* exists. As John Mayberry puts it,

> The set-theoretical axioms that sustain modern mathematics are self-evident in differing degrees. One of them – indeed, the most important of them, namely…the so-called axiom of infinity – has scarcely any claim to self-evidence at all.   [2000, 10]

Other axioms are even less evident. In the context of the other axioms, the Axiom of Replacement implies the existence of outrageously huge (but tiny for set theory!) infinite sets, such as a cardinal number, κ, larger than all f(i), where $f(0) = \aleph_0$ and $f(i + 1) = \aleph_{f(i)}$. f(0) has the size of the set of natural numbers, ω, while already f(1) has the "size" of the *ωth cardinal number*. (Actually, f(0) *is* ω, and similarly for f(1), as card(x) = x, for any cardinal, x.) That is a fast growing function! It is not "self-evident" that κ exists![9] As Thomas Forster writes, "advocates of the axiom scheme of replacement do not claim obviousness for their candidate…It is often said to be plausible, but even that is pushing it. 'Believable' would be more like it" [Forthcoming, 10].

Similarly, reasons to doubt Choice are well known. I will mention one here. In the context of the other axioms, Choice is equivalent to the claim

[9] George Boolos does not just deny that it is self-evident that κ exists. He denies that it exists. He writes, "Let me try to be as accurate, explicit, and forthright about my belief about the existence of κ as I can…I…think it probably doesn't exist.…I am also doubtful that anything could be provided that should be called a *reason* and that would settle the question" [1999, 121, italics in original]. One could in principle deny that self-evidence is closed under known entailment. But it is hard to see what the use of such a watered-down notion of self-evidence would be.

that every set is well-orderable—that is, totally orderable so that every non-empty subset of it contains a least element. So, Choice implies that the set of real numbers, R, is well-orderable. But the order in question clearly cannot be the usual one, since there is no least real number in the open subset (0, 1), which excludes 0 and 1. Moreover, it is consistent with standard set theory with Choice (if that is consistent!) that there is no definable well-order on R at all—that is, no well-order given by a formula, no matter how complex. Even if it is *true* that R is well-orderable, again, it is not self-evident that this is so! Forster says, "[w]ithout any doubt the most problematic axiom of set theory is the axiom of choice.... The current situation with AC is that the contestants have agreed to differ" [Forthcoming, 58].[10]

Similar remarks apply to other axioms, like the Powerset Axiom, which says that whenever a set, S, exists, the set of all of S's subsets does too. Speaking of a variety of axioms, Boolos writes:

> I am by no means convinced that any of the axioms of infinity, union, or power [set]...force themselves upon us or that all the axioms of replacement that we can comprehend do.... [T]hat there are doubts about the power set axiom is of course well known...[T]here is nothing *unclear* about the power set axiom....But it does not seem to me unreasonable to think that...it is not the case that for every set, there is a set of all its subsets.   [1999, 130–1, italics in original][11]

Of course, if standard axioms like Infinity, Replacement, Choice, and Powerset are not self-evident, then extensions of them, namely so-called *large cardinal axioms*, are not a fortiori—contra the rhetoric of some mathematicians. Woodin suggests that the fact that the Axiom of Constructibility, $V = L$ (which says that all sets are constructible, in a technical sense), implies that "large" large cardinals, like a Measurable Cardinal, do not exist constitutes a reductio ad absurdum of $V = L$—as if *that there is a Measurable Cardinal* were self-evident. He writes,

---

[10] It is sometimes suggested that all disagreements over Choice bottom out in disagreement over classical logic (with detractors rejecting classical logic in favor of intuitionistic logic). But, as Forster [Forthcoming, Ch. 7] illustrates, this is not so.

[11] An example of a version of set theory that lacks the Powerset Axiom, as well as instances of Replacement (and Comprehension, which I have not discussed), is Kripke-Platek set theory.

Godel's *Axiom of Constructibility*, V = L, provides a conception of the Universe of Sets which is perfectly concise modulo only large cardinal axioms which are strong axioms of infinity. However the axiom V = L limits the large cardinal axioms which can hold and so the axiom is *false*. [2010, 1, emphasis in original]

However, one set theorist's modus ponens is another's modus tollens! Ronald Jensen writes, "I personally find [V = L] a very attractive axiom" [1995, 398]. He elaborates:

[V=L] on the one hand and large cardinals and determinacy on the other embody two radically different conceptions of the universe of sets. How can these conceptions be justified? Most proponents of V = L and similar axioms support their belief with a mild version of Ockham's razor. L is adequate for all of mathematics; it gives clear answers to deep questions; it leads to interesting mathematics. Why should one assume more?....I do not understand...why a belief in the objective existence of sets obligates one to seek ever stronger existence postulates. Why isn't Platonism compatible with the mild form of Ockham's razor...?   [1995, 401][12]

---

[12] The standard narrative is that V = L is false, and "clearly" so. Maddy [1997, Pt II, § 4] contains a nice explication of the reasons advocates of this narrative supply. But the claim seems to have roughly the standing as the claim that, say, Timothy Williamson's E = K is false, and clearly so. Harvey Friedman writes, "[some s]et theorists say that V = L has implausible consequences....[They] claim to have a direct intuition which allows them to view these as so implausible that this provides 'evidence' against V = L. However, mathematicians [like me] disclaim such direct intuition about complicated sets of reals. Many say they have no direct intuition about all multivariate functions from N into N!" [2000]. Indeed, the early Godel (who discovered the axiom) appeared to favor V = L. He writes, "The proposition A (i.e., V = L) added as a new axiom seems to give a natural completion of the axioms of set theory, in so far as it determines the vague notion of an arbitrary infinite set in a definite way" [1990/1938, 27]. More recently, Keith Devlin writes, "What is my own view?...Currently I tend to favour [V=L]....At the moment I think I am in the majority of informed mathematicians, but the minority of set theorists..."[1981, 205]. W. V. O. Quine writes, "sentences such as the continuum hypothesis and the axiom of choice, which are independent of [standard] axioms, can...be submitted to the considerations of simplicity, economy, and naturalness that contribute to the molding of scientific theories generally. Such considerations support Gödel's axiom of constructibility, 'V = L'" [1990, 95]. Tatiana Arrigoni says, "I believe it perfectly in order to characterize...ZFC + V = L as intuitively plausible..." [2011, 355]. And Charles Pinter writes, "[T]here is a strong intuitive basis for considering L to be the class of all sets. By definition, L contains all the sets that are describable by a formula in the language of set theory. And there is no practical reason to admit sets which lack any description, for we would never make use of such sets. They would merely sit there and muddy the waters." [2014/1971, 227]

Again, even if Woodin is right and Jensen is wrong, it would be pointless to deem the proposition *that there exists a Measurable Cardinal* self-evident—when a world expert suggests that it is false, and when disagreement over it need not turn on any outstanding conjectures.

2.4 Plausibility and Disagreement

So, if our *non-trivial* mathematical beliefs have better claim to being a priori justified than our non-trivial moral beliefs, realistically construed, this is not because the former are provable from self-evident or intuitively obvious axioms, while the latter are not—contrary to a cartoon of mathematics which prevails in many quarters. It is not, that is, because:

> mathematics begins with a small number of shared, self-evident assumptions, while morality begins with a large number of inter-connected assumptions...all of which sound reasonable to the assumption-maker and precious few of which are truly self-evident.   [Greene, 2013, 184–5]

But, as I indicated above, few philosophers of mathematics would have suggested otherwise.[13] Already in his [1973/1907] work, Bertrand Russell spoke of the *inductive method* for discovering the axioms of mathematics. He writes:

> We tend to believe the premises because we can see that their consequences are true, instead of believing the consequences because we know the premises....But the inferring of premises from consequences is the essence of induction; thus the method in investigating the principles of mathematics is really an inductive method, and is substantially the same as the method of discovering general laws in any other science.
>
> [1973/1907, 273–4][14]

---

[13] See Forster [Forthcoming], Fraenkel, Bar-Hillel, and Levy [1973], Maddy [1988a and 1988b], Shapiro [2009] for more on the untenability of this simplistic picture of mathematics.

[14] See also Whitehead and Russell [1997]: "The reason for accepting an axiom, as for accepting any other proposition, is always largely inductive, namely that many propositions which are nearly indubitable can be deduced from it, and that no equally plausible way is known by which these propositions could be true if the axiom were false, and nothing which is probably false can be deduced from it [59]."

Godel elaborates on Russell's method this way:

> [Russell] compares the axioms of…mathematics with the laws of nature and logical evidence with sense perception, so that the axioms need not…be evident in themselves, but rather their justification lies (exactly as in physics) in the fact that they make it possible for these 'sense perceptions' to be deduced.…I think that…this view has been largely justified by subsequent developments, and it is to be expected that it will be still more so in the future.   [1990/1944, 121][15]

Russell's inductive method is reminiscent of John Rawls's method of *reflective equilibrium*—that is, the method of "testing theories against judgments about particular cases, but also testing judgments about particular cases against theories, until equilibrium is achieved" [Blackburn 2008, 312].[16] We identify propositions that we deem plausible—though rarely self-evident—and seek principles, "axioms," which systematize them. Such principles may pressure us to reject the propositions with which we began as we seek optimum harmony between the two. Proof can play a role in the process.

---

[15] Or, again: "The so-called logical or set-theoretical 'foundation' for number-theory or of any other well established mathematical theory, is explanatory, rather than really foundational, exactly as in physics where the actual function of axioms is to *explain* the phenomena described by the theorems of this system rather than to provide a genuine 'foundation' for such theorems" [quoted in Lakatos 1976, 204, italics in original]. Imre Lakotos follows this with the following from Hermann Weyl: "A truly realistic mathematics should be conceived, in line with physics, as a branch of the theoretical construction of the one real world, and should adopt the same sober and cautious attitude toward hypothetic extensions of its foundations as exhibited by physics" [quoted in Lakotos 1976, 204]. Although the philosophical and mathematical literature on "inductive" considerations (in Russell's sense) focuses on new axioms (e.g., a Measurable Cardinal versus V = L), Russell is clear in his [1973/1907] work that he even takes axioms of arithmetic to be supported inductively. Similarly, Donald Martin observes, "[E]ven for some of the ZFC axioms, the intrinsic evidence is not the main evidence. Consider…the Axiom of Infinity. There is certainly intrinsic evidence for it. But there is intrinsic evidence against it that is at least as compelling: the…idea that there can be no completed infinite totalities, for example.…Woodin (1988) might be taken as making a similar point even about Peano arithmetic…" [1998, 229—30]

[16] See Clarke-Doane [2014, § 1]. Rawls himself credits the method of reflective equilibrium to Goodman [1955, 63–4], characterizing the method somewhat differently in different places. One of his statements is the following. "People have considered judgments [about morality] at all levels of generality, from those about particular situations and institutions up through broad standards and first principles to formal and abstract conditions on moral conceptions. One tries…to fit these various convictions into one coherent scheme, each considered judgment whatever its level having a certain initial credibility. By dropping and revising some, by reformulating and expanding others, one supposes that a systematic organization can be found. Although in order to get started various judgments are viewed as firm enough to be taken provisionally as fixed points, there are no judgments of any level of generality that are in principle immune to revision" [1974, 8]. See also Rawls [1971] and Parsons [2009, 324], who mentions reflective equilibrium when discussing the justification of our mathematical beliefs as well. But Parsons appears to view its role as quite limited.

For example, proof tells us that Choice implies a range of plausible propositions, while apparently refuting some as well. It implies, for example, that union of countably many countable sets is countable, and that any two sets either have the same cardinality or one has greater cardinality—that is, the Principle of Cardinal Comparability (PCC). More dubiously, it implies the so-called Banach–Tarski Paradox and that the real numbers are well-orderable. But whether such results show that we ought to endorse or reject Choice is left open. That depends on which alternative would facilitate equilibrium among our beliefs.[17]

Perhaps, then, even if our mathematical beliefs are not provable from self-evident axioms, an analog to Rachels's contrast still holds. It may be that the mathematical "data points" on which the method of reflective equilibrium—as opposed to the method of mathematical proof—operates are epistemically privileged in a sense in which corresponding moral propositions are not. Perhaps the mathematical propositions that we deem plausible (which, again, are not generally axioms) have an epistemic standing that the moral propositions that we deem likewise lack. (I will henceforth call propositions that we deem plausible "plausible propositions.")

But what kind of epistemic standing could, for example, the Principle of Cardinality Comparison (PCC) enjoy that, say, the principle that slavery is wrong does not? More to the point, what could our *evidence* be that PCC enjoys that standing while the proposition that slavery is wrong does not?

It might be thought that our evidence is that everyone believes PCC, but there are those among us who do not believe that slavery is wrong. This cannot be right, however. To even understand PCC, one needs acquaintance with set theory, a field that vanishingly few people have ever even heard of. A slightly better proposal would be that everyone who understands plausible mathematical propositions, like PCC, believes them, but this is not so in the moral case. But this cannot be right either. There are people who understand plausible mathematical propositions but do not believe them either. Mathematical fictionalists do not believe any plausible mathematical propositions, because they do not believe any mathematical propositions *at all* (see Section 1.2). (More exactly, fictionalists do not believe any atomic or existentially quantified mathematical propositions, and believe that all universally quantified mathematical propositions are vacuous.)

It might be objected that fictionalists are beside the point. For every area, F, be it physics, philosophy, or logic, there are F-fictionalists—that is,

---

[17] This runs contrary to the contrast between our moral and mathematical beliefs suggested by Kelly and McGrath [2010, 341].

philosophers who understand F-claims as well as the rest of us but take them to involve commitments which are not satisfied. But that does not show that no kinds of claim possess a privileged epistemic standing. What matters is that, *bracketing* fictionalists, everyone believes plausible mathematical propositions, but this is not so of plausible moral propositions. However, even if we take fictionalists to be irrelevant in this context (which strikes me as a difficult posture to defend), there are those who deny plausible mathematical propositions, though they are not mathematical fictionalists. For example, skeptics about Choice, such as Michael Potter [2004, Pt IV, § 14], will tend to be skeptics about PCC, since Choice and PCC are actually equivalent in the context of the other axioms.

It might be thought that the picture looks different if we focus on ordinary branches of mathematics, as opposed to set theory. As I said at the outset, set theory is a fringe area of modern mathematics. What about core areas which are commonly thought to be objective?

Various theorists doubt plausible principles from core areas too. For instance, the basic principle of the calculus, the Least Upper Bound Axiom (which says that every non-empty set of real numbers with an upper bound has a least upper bound), is plausible if anything is. But Hermann Weyl maintains that "[i]t will be recognized...that in any wording [the Least Upper Bound Axiom] is false" (quoted in Kilmister [1980, 157]). Surely, though, bracketing fictionalists, at least *arithmetic* is off limits from serious skepticism? On the contrary, the Princeton mathematician, Edward Nelson, rejects instances of Induction (which says that if 0 has a property, F, and if n+1 has F whenever n has it, then all natural numbers have F) [1986, Introduction].[18] Doron Zielberger goes further. He writes, "I am a platonist...[but] I deny even the...Peano axiom that every integer has a successor..." [2004, 32–3]. Actually, Nelson sometimes appears to reject the Successor Axiom as well, at least in connection with "actual" (or "genetic") numbers ([1986, 176]). Elsewhere Nelson elaborates on his position as follows (where Q is Robinson Arithmetic, that is, Peano Arithmetic minus the Induction Schema):

To avoid vagueness, let Q^* be Q with the usual relativization schemata adjoined. Construct a formal system F by adjoining an unary predicate symbol ψ, the axiom $\psi(0)$, and the rule of inference: from $\psi(a)$ infer $\psi(Sa)$ (for any term a). I think this is an adequate formalization of the concept of

---

[18] Note that Weyl [1918] and Nelson [1986] accept classical logic—though Nelson's discussion of the Successor Axiom seems to show that he must jettison the Deduction Theorem. See the quotations to follow.

an "actual number." Is $\psi(80\wedge5000)$ a theorem of F? I see no reason to believe so. Of course, one can arithmetize F in various theories, even Q^*, and prove a formula $\exists p[p$ is an arithmetized proof in F of '$\psi(80\wedge5000)$'], but to conclude from this that there is a proof in F itself of $\psi(80\wedge5000)$ appears to me to be unjustified. Contrast F with the theory T obtained by adjoining to Q^* a unary predicate symbol ϕ and the two axioms $\phi(0)$ and $\phi(0)\&\forall x'[\phi(x')\rightarrow\phi(Sx')]\rightarrow\phi(x)$. Then one can easily prove in T $\phi(80\wedge5000)$ or even $\phi(80\wedge5000\ldots\wedge5000)$. The ellipsis means that the iterated exponential term is actually written down.

He continues,

Q^* proves that addition and multiplication are total, but does not prove that exponentiation is total. The situation is different with F and ψ: a rule of inference is far more restrictive than an implication, and F does [does not even] prove that if $\psi(x)$ then $\psi(Sx)$.[19]

Harvey Friedman laments, "I have seen some ... go so far as to challenge the existence of 2^{100} ... "   [2002, 4].

2.5 Extent of Disagreement

Perhaps universal assent was too high a bar—even bracketing fictionalists. There are always heretics, even brilliant ones. That does not show that no kinds of proposition possess a privileged epistemic status. What matters is the *extent* of disagreement. As Brian Leiter writes:

[P]ersistent disagreement on foundational questions ... distinguishes moral theory from inquiry in ... mathematics, certainly in degree. [2009, 1]

Likewise, Tristram McPherson writes:

Presumably our access to mathematics ... comes via intuitive reflection. But if our mode of epistemological access to morality and to mathematics is identical, why is there a striking contrast between our track records of theoretical progress in these fields?   [Forthcoming a, 8]

---

[19] See the full discussion here: https://mathoverflow.net/questions/142669/illustrating-edward-nelsons-worldview-with-nonstandard-models-of-arithmetic.

The above discussion illustrates that the notion of extent of disagreement is actually ambiguous. First, disagreement from an area, F, may be *propositions-widespread*, in that, for many F-propositions, Q, there is a pair of people, P, such that P disagrees with respect to Q. Actually, since disagreement over Q always translates into disagreement over ~Q, ~~Q, and so on ad infinitum, we should require that there are many *kinds* of F-propositions, G, such that there is a pair of people, P, and P disagrees with respect to a G-proposition, Q. The above discussion suggests that mathematical disagreement is no less propositions-widespread than moral disagreement. Again, there is disagreement among non-fictionalists over everything from set theory to analysis to arithmetic. And if we include fictionalists, there is disagreement over literally every atomic or existentially quantified mathematical proposition at all. But there is another way to understand "extent" of disagreement. Perhaps what matters is whether mathematical disagreement is less *people-widespread* than moral disagreement—where this means, roughly, that there are fewer pairs of people who disagree over some mathematical propositions than there are pairs of people who disagree over some moral propositions.

Who are the relevant people? There is a cheap sense in which there are fewer pairs of people who disagree over mathematical propositions. It is the sense in which there are fewer pairs of people who disagree over epistemo-logical internalism. There are very few mathematicians, and dramatically fewer people who have seriously considered the question of what axioms are true. Mathematicians are overwhelmingly focused on questions of *logic*—that is, what is true *if the axioms are true*. As George Kriesel puts it, philosophical "foundations provide reasons *for* axioms, *practice* is concerned with deductions from axioms" [1967, 191, emphasis in original]. Ask a typical mathematician whether the axioms themselves are true, and you are unlikely to get a stable answer.[20] Actually, even this is an overstatement. Mathematicians are overwhelmingly focused on the even more conditional question of what is true if the axioms are, *assuming classical logic*. Ask a mathematician whether classical logic is the "right" logic, or even whether

---

[20] See, again, Kriesel [1967] for a helpful overview of philosophers/foundations theorists, mathematicians, and "the practical man's" diverse aims and knowledge. Hans Reichenbach notes that "The philosophic analysis of the meaning and significance of scientific statements can almost hinder the processes of scientific research and paralyze the pioneering spirit, which would lack the courage to walk new paths without a certain amount of irresponsibility" [1956/1927, xiii].

there *is* a right logic (or even what that question means), and you are unlikely to get a stable answer.[21] This is no criticism of mathematicians, contra Frege [1980/1884].[22] As Easwaran [2008] notes, nobody would ever prove any theorems if they had first to settle the question of what axioms are true! Mathematicians simply avoid the question. Trivially, then, there are many fewer pairs of people who disagree over mathematical propositions, because there are many fewer people who have a serious view as to what mathematical propositions are true.

It might be objected that, on the contrary, we all have at least arithmetic views. We believe that 2 + 2 = 4, that there are infinitely many prime numbers, and so forth. But this is questionable. What is hard to deny is that we believe things which we reliably convey (perhaps only pragmatically) with sentences like "2 + 2 = 4."[23] Realistically construed, however, such a sentence is about *numbers*. It says that 2 bears the plus relation to itself and to 4. Do most people really believe that? It seems at least as likely that they believe such *first-order logical truths* as that if there are "exactly two" apples on the table, and "exactly two" apples on the chair, and nothing on the table is on the chair, then there are "exactly four" apples on the table or on the chair (where the phrases "exactly two" and "exactly four" here are definable in terms of ordinary quantifiers plus identity). Similarly, do most really people believe that, among the furniture of reality, including planets, particles, and passenger trains, are numbers, and infinitely many prime ones? It seems as likely that they believe something more along the lines of the following. If standard mathematical principles are true (whatever those are),

---

[21] Since a mathematical proof purporting to show that S follows from T uses classical logic, if it fails to be, say, intuitionistically valid, then it will do nothing to convince an intuitionist that S follows from T. One might think that it at least shows that S follows from T *in classical logic*. But even the truth of *this* claim can depend on what logic one uses to check. In general, whether *S follows from T in logic L* is itself dependent on the background logic (Shapiro [2014], ch. 7]). The if-thenist idea that there are only system-relative facts about what follows from what "all the way down" is dubiously intelligible.

[22] Gottlob Frege famously writes, "[Foundational questions] catch...mathematicians, or most of them, without any satisfactory answer. Yet is it not a scandal that our science should be so unclear about the first and foremost among its objects [as, e.g., the nature of the number 1], and one which is apparently so simple? Small hope, then, that we shall be able to say what number is. If a concept fundamental to a mighty science gives rise to difficulties, then it is surely an imperative task to investigate it more closely until those difficulties are overcome....Many...will be sure to think this is not worth the trouble....The result is that we still rest content with the crudest views" [2003/1884, 11].

[23] See Edidin [1995] for the view that atomic mathematical sentences are literally false, but we pragmatically convey truths with them.

then there are infinitely many prime numbers. This conditional is consistent with the non-existence of prime numbers.

Even if it were granted that all non-fictionalists believe the likes of $2 + 2 = 4$, however, this would still not establish a lack of parity with the moral cases. $2 + 2 = 4$ is among the most rudimentary mathematical propositions that there are. It is also difficult to think of any non-error theorist who denies that it is sometimes morally permissible for some people to stand, or—to invoke a favorite from the ethics literature—that it is wrong to burn babies just for the fun of it.

It is true that the above account of the contents of our arithmetic beliefs will not suffice to explain all apparent arithmetic convergence.[24] Consider, for example, the theorem that Pell's Equation, $x^2 - 2y^2 = 1$, has infinitely-many solutions. Different historical figures appear to have independently arrived this result again and again, despite their mutual isolation. Whatever the precise content of the theorems they proved, it cannot easily be identified with a logical truth. Unlike "$2 + 2 = 4$," there are no surrogate logical truths in this case. And, unlike contemporary mathematicians, premodern mathematicians rarely proceeded from explicit axioms.

However, there are also interesting examples of moral propositions which are not only widely accepted by non-error theorists, but which show up again and again across cultures. The largest cross-cultural survey conducted to date finds seven such propositions, concerning family, group, reciprocity, bravery, respect, fairness, and property [Curry et al. 2019]. Of course, it is hard to know what to make of such findings. How to translate group norms into regimented English, and, hence, how to measure agreement, is vexed. But the translation of arithmetic theorems into regimented English is difficult too. For instance, Euclid states his theorems as recipes for construction, rather than propositions about numbers. Did he believe what we do?

It would not help to complain that I am talking about a philosophical interpretation of "$2 + 2 = 4$," while what matters is the distribution of opinion on the question of whether $2 + 2 = 4$. Our question is the relative epistemological standing of moral and mathematical propositions, *realistically construed*—not construed under some interpretation or another. Indeed, the question of realism for an area, F, would be entirely trivialized if it were taken to be whether people believe (atomic) F-sentences under some construal or another. *Every (first-order) consistent sentence whatever is true under*

---

[24] Thanks to Conor Mayo-Wilson for discussion of the following points.

some interpretation or another, by the Completeness Theorem. Surely, one does not count as a realist about *everything* on account of this truism![25]

What matters, then, is whether there are fewer pairs of people who disagree over *non-trivial* mathematical propositions—where this is not just a matter of the absolute number of people who disagree. Roughly, we would like to know whether the *proportion* of those with views as to what non-trivial mathematical propositions are true who disagree is less than the proportion of those with views as to what non-trivial moral propositions are true who disagree. But even this is too crude a metric. For all that has been said, most people with views as to what non-trivial mathematical propositions are true may believe standard axioms simply because they have been taught *that standard axioms are "the axioms."* As D. A. Martin puts it, "[f]or individual mathematicians, acceptance of an axiom is probably often the result of nothing more than knowing that it is a standard axiom" [1998, 218]. If so, then the fact that most people with views as to what mathematical propositions are true agree that standard axioms are would hardly be evidence that those propositions possess a privileged status that plausible moral ones do not.

Let us imagine that the situation were reversed—that is, that there was virtual unanimity among non-error theorists as to the truth of non-trivial moral propositions. But among those with views on plausible moral propositions, everyone went to "morality school," and, on the first day, were given the "morality axioms." (Morality school then mostly consisted in proving moral theorems from the morality axioms.) Finally, vanishingly few morality school students were ever so much as made aware of arguments or evidence bearing on the morality axioms, and hardly anyone at morality school even thought twice about them.[26] Then, surely the said convergence would not be evidence that plausible moral propositions possess a privileged epistemic status.

What, then, would constitute such evidence? Perhaps convergence among the *knowledgeable*. If the proportion of those who were knowledgeable of arguments and evidence bearing on such principles as PCC, or the Least Upper Bound Axiom, and disagreed was much less than the proportion

---

[25] The Completeness Theorem assumes some basic set theory. But that is irrelevant. The point is that a realist about basic set theory does not automatically count as a realist about any (first-order) consistent theory, just because one can cook up a set-theoretic interpretation under which it is true.

[26] Barbara Herman suggests that there is a sense in which we *do* go to morality school. See her [2007, ch. 6] work.

of those who were knowledgeable of arguments and evidence bearing on such propositions as that slavery is wrong, or that capital punishment is unjust, and disagreed, then perhaps this would be evidence that plausible mathematical propositions enjoy a privileged epistemic status that plausible moral propositions do not. But, on the contrary, *among those (vanishingly few) who are knowledgeable of pertinent arguments and evidence*, there is notorious disagreement in the mathematical case. Fraenkel, Bar-Hillel and Levy note that there is "far-going and surprising divergence of opinions and conceptions of the most fundamental mathematical notions, such as set and number" among experts. [1973, 14] And John Bell and Geoffrey Hellman write,

> Contrary to the popular (mis)conception of mathematics as a cut-and-dried body of universally agreed upon truths…as soon as one examines the foundations of mathematics one encounters divergences of viewpoint… that can easily remind one of religious, schismatic controversy.   [2006, 64]

Bell and Hellman do add the following qualification:

> While there is indeed universal agreement on a substantial body of mathematical results…as soon as one asks questions concerning fundamentals, such as… "What axioms can we accept as unproblematic?" … we find we have entered a mine-field of contentiousness   [2006, 64].

But if such fundamentals are contentious, then the "substantial body of mathematical results" to which Bell and Hellman allude is apparently limited to *logical truths* of the form, *if T, then S*, where T is the conjunction of the finitely-many axioms used in the proof of S. Again, we could achieve like agreement over such moral logical truths. We could regiment the various moral theories in classical first-order logic and prove "moral theorems." Forster characterizes the situation baldly:

> [F]or people who want to think of foundational issues as resolved… [standard axioms provide] an excuse for them not to think about [them] any longer. It's a bit like the role of the Church in Medieval Europe: it keeps a lid on things that really need lids. Let the masses believe in [standard] set theory. To misquote Chesterton "If people stop believing in set theory, they won't believe nothing, they'll believe anything!"   [Forthcoming, 15]

So, contra Leiter and McPherson, there does not seem to be an *epistemologically important* sense in which morality is more controversial than mathematics—even though there are obviously dramatically more people who disagree over moral propositions.

2.6 Philosophical Corruption

It might be thought that there is something fishy about the above argument. Do not arguments and evidence bearing on plausible mathematical propositions take things which are obvious and *make them* controversial? Such arguments are a corrupting influence, as it were. It might be thought that the only people who are skeptical of Induction have been introduced to principles like *predicativism* (which says that it is not coherent to define an object in terms of a totality to which it belongs) or *ultrafinitism* (which says that not even large finite numbers exist). But these principles stand to philosophical principles like *epistemic closure* as Nelson's and Zielberger's views stand to skepticism. Epistemic closure says that if X knows that P, and X knows that P implies that Q, then X knows that Q. If epistemic closure is true, and we know that we have hands, then we must know that we are not brains in vats. But, we might nervously query, do we know this? Such principles lead us to question what was obvious to start with. Russell quips:

> My desire and wish is that the things I start with should be so obvious that you wonder why I spend my time stating them. This is what I aim at because the point of philosophy is to start with something so simple as not to seem worth stating, and to end with something so paradoxical that no one will believe it.   [1918, 514]

Perhaps, then, what distinguishes plausible mathematical propositions from plausible moral ones is that anyone who considers the former will at least find them to be *initially credible*—even if philosophical arguments lead one astray. The same is not true of plausible moral propositions.

There are three problems with this suggestion, however. First, *who knows* what moral propositions we would find credible if all philosophical considerations were bracketed. (Of course, if moral propositions count as philosophical, then we would find no such propositions credible.) What would it even mean to have a serious view as to whether, for example, Osama bin Laden was morally blameworthy for killing 2,996 people *bracketing* the question of whether we have free will, whether God exists, and whether moral relativism is true?

Second, to the extent that we can make sense of the analogous question in the mathematical case, it is doubtful that there is convergence on a significant range of credibility judgments of interest. Again, when Weyl proclaimed that "[i]t will be recognized…that in any wording [the Least Upper Bound Axiom] is false," he did not seem to be registering theoretical doubt. Or consider the intuitions that are traded in discussions of the backbone to standard set theory, the Axiom of Foundation (which says that every set occurs at some level of the cumulative hierarchy). The debate over Foundation has virtually no consequence for mathematics generally (Kunen [1980, ch. 3], Fraenkel, Bar-Hillel, and Levy [1973, 87]). The central question seems to be whether it is credible that there are sets that contain themselves or whether there are sets which have infinitely descending chains of membership. The standard view, if there is one, is that it is not (Boolos [1971, 491, [Maddy, 1988a]). Such sets are pathological. But some have a very different view. For instance, Adam Rieger writes,

A theory of sets should…be answerable to our informal concept of set as completely arbitrary collection, as well as to the needs of mathematicians…. Only a non-well-founded theory [that is, a theory that is inconsistent with Foundation] can…be shown to modify the naive conception as much as, but no more than, is required….   [2011, 17–18][27]

The dispute over the Axiom of Foundation seems to have more to do with what Jensen calls "deeply rooted differences in mathematical taste" than with philosophical priming [1995, 401].[28]

Finally, even if there is significant convergence on initial credibility judgments in mathematics, but not in morality, it is hard to see how this could

---

[27] See also Azcel [1988, Introduction]. Advocates of the so-called logical conception of set, such as Quine [1937 and 1969], also reject the Axiom of Foundation. Quine's New Foundations for mathematical logic (NF) proves the existence of a universal set, which contains itself.

[28] There is some evidence that divergent conceptions even of the natural numbers preexist axiomatic mathematics. Relaford-Doyle and Núñez [2018, 235] argue that "even highly educated people often seem to rely on conceptualizations of natural number that are different from, and even at odds with, the characterization given by the Dedekind-Peano axioms." And Pantsar writes of Mayan mathematics that "their arithmetic seems…to have been remarkably different from ours. While they could calculate with extremely high precision, they did not prove general truths about numerosities. Moreover, they did not seem to have a concept of the infinity of the them" [2014, 4218]. He concludes that "differences in the more developed theories suggest that the shared initial concept of discrete numerosity underdetrmines the development of arithmetic. It seems that we shared the concept of numerosity with the Mayans when it came to calculations, but in the end developed arithmetic differently" [2014, 4219]. See Section 5.8.

matter. Again, mathematicians' initial credibility judgments may be "the result of nothing more than knowing [what] is a standard axiom." Indeed, even if those judgments were biologically innate, it is hard to see how this could matter. Suppose that we were biologically programmed to find the proposition that retribution is morally justified initially credible. Who would seriously allege that *this fact* does anything to show that that the proposition in question enjoys a privileged epistemic standing—given, that is, that on reflection the proposition is evidently false or even just questionable?

2.7 The Concept of Set

I have argued that if our mathematical beliefs have any better claim to being (defeasibly) a priori justified than our moral beliefs, this is not because they (or those from which they inductively follow, in Russell's sense) have better claim to being provable, self-evident, plausible, or even initially credible, in a way that our moral beliefs are not—contrary to what is widely alleged. Is there any other way to argue that our mathematical beliefs have better claim to being a priori justified than our moral beliefs? It might be thought that there is the following way.[29] One can argue that our mathematical beliefs are *analytic,* while our moral beliefs are not. Perhaps, for instance, it is just "part of the concept of set" that plausible set-theoretic principles hold, while nothing analogous can be said of plausible moral principles. And maybe if a proposition is part of a concept in the appealed to sense, then those of us with the concept count as being at least (defeasibly) a priori justified in believing the proposition. Following Boghossian [2003]), we might call propositions with this status *epistemically*, rather than metaphysically, analytic, so as to avoid any suggestion of the truth in virtue of meaning to which Quine [1951b] objected.

But appeal to epistemic analyticity is of little use in this context. First, it is hard to think of a non-question-begging argument that any set-theoretic principle of interest is just "part of the concept of set," given that some theorist actually denies it.[30] Consider, again, the Axiom of Foundation. This is often supposed to be a prime example of a principle that follows from the concept of set (Boolos [1971, 498], Shoenfield [1977, 327]). It is just part of

---

[29] Thanks to Paul Boghossian for pressing me on this. See also Wright and Hale [2002] and Burgess [2005, Pt III].
[30] See Williamson [2006] for a cognate argument.

what we mean by "set" that every set is formed at some stage of a transfinite generation process via the power-set and union operations, beginning with the empty set— so no set contains itself, and there are no infinitely descending chains of membership. But far from being beyond dispute, many think that this claim is not even *coherent*! What, after all, does "formation" and "generation" even *mean* when applied to abstract entities like (pure) sets (Ferrier [Forthcoming], Potter [2004, § 3.3])?[31]

Of course, if a principal that is supposed to be definitive of the concept of set, like Foundation, is not beyond dispute, then others—such as CPP—are not a fortiori. Why should any two sets be comparable by cardinality just because sets are formed at stages as above?

An advocate of the epistemic analyticity of Foundation might respond that it was never part of the view that no one denies epistemically analytic propositions. However, the problem is to explain why we should *believe* that Foundation is epistemically analytic despite fervent opposition by conceptually competent theorists, while *denying* the epistemic analyticity of any moral principle of interest. After all, some moral realists have argued for a precisely analogous view in connection with "a battery of substantive moral propositions" (Cuneo and Shafer-Landau [2014]). If the reason to deny the epistemic analyticity of such moral principles is not the persistent opposition to them by obviously conceptually competent people, then what is it?[32]

Second, even if we grant that plausible set-theoretic principles are epistemically analytic while no moral principles are, this would just relocate the epistemic problem. If we were worried that some sets fail to be comparable in cardinality, say, then under the assumption that it is just part of the concept of set that all sets are, we should just worry *that our concept of set is not satisfied*. (One does not get to avoid speculative metaphysics by simply enriching one's concepts!) Maybe instead of sets, there are only shmets— where shmets are just like sets except that some shmets fail to be so comparable. In general, if we grant that there is a (non-empty) objective set-like reality, then there is a translation between the worry that sets might fail to be as we take them to be, assuming that no set-theoretic principles are

---

[31] Again, Rieger complains, "[The iterative concept of set] does not embody a philosophically coherent notion of set. There is a coherent constructivist position....There is also a coherent anti-constructivist position....But [the iterative concept of set] is an uneasy compromise between these two: it pays lip-service to constructivism without...meaning it..." [2011, 17–18].

[32] In Chapter 6 I will suggest that what I call *practical* questions may be essentially contestable in a way that no other kinds of questions are. But these do not include moral questions, realistically, or even factually, construed. On realism, see Section 1.2.

epistemically analytic, and the worry that there might be shmets instead of sets, assuming that all such principles are (Clarke-Doane [2014, § III]). Of course, this translation does not preserve meaning, since it does not even preserve reference. It preserves the epistemic mystery—or one much like it. Even if there were a non-question-begging argument that set-theoretic principles have better claim to being epistemically analytic than moral principles, it is hard to see how this would show that our set-theoretic beliefs have better claim to being a priori justified than our moral beliefs.

If we could argue that every "consistent" concept of set is satisfied, then we might be able to rule out the worry that our concept of set is not. I will consider such an argument in detail in Sections 5.9 and 6.1. However, to endorse such an argument would be give up on the whole project of trying to determine what mathematical axioms are true. It would be to embrace the most radical form of set-theoretic anti-objectivism (in the sense of Section 1.6), according to which every consistent mathematical sentence which is not a logical truth is like the Parallel Postulate (even the one that codes the claim that arithmetic is consistent!). As I said in Section 1.6, this position is practically indistinguishable from the most uncompromising relativism.

Is there any other argument for the view that our mathematical beliefs have better claim to being a priori justified that is worth taking seriously? I am not aware of one. Of course, it is possible that plausible mathematical propositions do possess a privileged epistemic standing that plausible moral propositions do not. Perhaps the former really are epistemically analytic. Indeed, perhaps anyone—or anyone who understands—the former is epistemically obligated to believe them, or find them plausible or initially credible, while the same is not true of the latter. But absent some evidence that this is so—such as the fact that plausible mathematical, but not moral, propositions are in fact deemed initially credible by all who understand them—this suggestion just blatantly assumes what is in question. It assumes that plausible mathematical propositions possess a privileged epistemic standing that plausible moral propositions do not.

2.8 Error Theory

I have argued that our mathematical beliefs have no better claim to being a priori justified than our moral beliefs, realistically construed. In particular, they have no better claim to being provable, self-evident, plausible, initially credible, or even epistemically analytic than our moral beliefs, so construed—contrary to what is widely maintained. But it will surely be responded that this conclusion is counterintuitive and goes against nearly all writing to date comparing morality to mathematics. If it is true, then what explains such strong appearances to the contrary?

Three factors explain them. First, there is an understandable, but widespread, confusion between arithmetic and (first-order) logic. We folk agree, and find it hard to deny, that two apples and another two make four apples in all. We take ourselves to thereby agree that $2 + 2 = 4$, realistically construed. But in this we err. Because the relevant sentences are intimately related, however, those of us not versed in the distinction naturally confuse the two.[33]

Second, there is a widespread conflation of two senses of "proof"—one logical, and the other epistemic. To prove something in the *logical* sense is to show that, given agreement over logic, if the axioms are true, then so too is the claim proved (as described in Section 2.2). By contrast, to prove something in the *epistemic* sense is to demonstrate that it is true (in some contextually specified sense of "demonstrate"). Thanks to the widespread use of logical proof in mathematics, there is a sense in which it has a method. But it is a method for proving things in the logical sense—that is, for demonstrating what is true if the axioms are—and, strictly speaking, it only does this in a context where there is agreement over logic. Moreover, *this method is available to us all*, whether mathematicians, ethicists, or astrologists. This fact should be of little consolation to ethicists because what is of ultimate *moral* concern is epistemic, not logical, proof. (Who cares what follows from arbitrary moral principles?) When it comes to determining what *non-logical* claims are true (whether mathematical or moral), however, we seem to be limited to applying the method of reflective equilibrium to propositions which strike us as true but are not beyond question, and which commit us to controversial, broadly philosophical, theses.[34] So, the *focus* of mathematicians is clearly different from that of ethicists (roughly, mathematicians want to know what is true if the axioms

---

[33] I illustrate the way in which they are related in Section 5.6. (Again, I do not claim that *all* rudimentary claims of everyday applied arithmetic amount to (first-order) logical truths. For instance, the claim that we cannot tile a floor in such and such a way, or that the number of tiles will be prime, has no first-order logical surrogate.)

[34] And, in a context in which the logic is in doubt, we seem limited to this method in determining what logical claims are true. Goodman [1955, 63–4], to which Rawls credits the method of reflective equilibrium, was focused on the case of logic, not mathematics or morality.

are true, while ethicists want to know what axioms are true). But the methods available to them are in principle the same.[35]

Finally, there are key differences between the *sociology* of mathematics and morality. I will mention four. (1) There is more *unanimity in mathematical practice*. The mathematical, and generally scientific, community has decided to *use*, for example, the Least Upper Bound Axiom, even if the case for its truth, realistically construed, is contested by predicativists in the tradition of Weyl [1918]. Again, most practicing mathematicians, let alone empirical scientists, simply do not see it as part of their job description to take a stand on the legitimacy of the impredicative definitions which they routinely use. This is as it should be, for reasons broached in Section 2.5.

(2) Moral disagreement tracks with familial and communal forces, while mathematical disagreement does not. Hence, John Mackie writes,

> [T]he actual variations in the moral codes are more readily explained by the hypothesis that they reflect ways of life than by the hypothesis that they express perceptions, most of them seriously inadequate and badly distorted, of objective values.... [1977, 37]

Mathematical disagreement does not track with familial or communal forces for the mundane reason that we are not introduced to the likes of the Least Upper Bound Axiom by our families or communities. But mathematical disagreement *does* seem to track with social forces. As the Fields medalist, Paul Cohen, observed, "the attitudes that people profess towards the foundations [that is, toward the question of what axioms are true] seem to be greatly influenced by their training and their environment" [1971, 10]. This is just the kind of contingency on which the epistemology of disagreement has focused. It is like the observation that had we gone to a different graduate school, we would have had different philosophical beliefs ([Cohen 2000, 18]).[36]

(3) Much more *hangs* on moral disagreements than hangs on mathematical ones. Couples break up on account of disagreements over moral principles. But typical mathematical disagreement is paradigmatically academic, like

---

[35] For more on this difference, see Section 6.7.
[36] Moreover, even if moral disagreement did track with social forces in some important way in which mathematical disagreement did not, it is hard to see how this would show that our moral beliefs have less claim to being (defeasibly) a priori justified than our mathematical beliefs. It might *undermine* those beliefs, if we became aware it. But that is a different point—one which I explore at length in Chapter 5.

typical (non-evaluative) philosophical disagreement—especially in light of the widespread agreement in mathematical practice mentioned above.[37]

(4) Finally, moral disagreement occurs among people of all levels of education and intelligence, and is peculiarly affected by religious, political, and emotional factors. No doubt mathematical disagreement can reflect such factors too. Fraenkel, Bar-Hillel, and Levy note of disagreement surrounding the Axiom of Choice that "both the positive and the negative attitude towards our axiom are *far more strongly influenced by emotional or practical reasons than by arguments or principles*" [1973, 81, italics in original]. But, given that moral disagreement is particularly sensitive to such factors, it is no wonder that it can appear especially intractable.

Ironically, however, this contrast *bolsters* the conclusion that our mathematical beliefs have no better claim to being a priori justified than our moral beliefs. Unlike moral disagreement, mathematical disagreement cannot be explained away as reflecting those distorting influences. By contrast, it is often argued that much, maybe even all, moral disagreement can be dismissed as stemming from ignorance, irrationality, religious dogma, and so forth. Derek Parfit writes,

> Belief in God, or in many gods, prevented the free development of moral reasoning. Disbelief in God, openly admitted by a majority, is a recent event, not yet completed. Because this event is so recent, Non-Religious Ethics is at a very early stage. We cannot yet predict whether, as in Mathematics [*sic*], we will all reach agreement. Since we cannot know how Ethics will develop, it is not irrational to have high hopes.   [1984, 454]

I take no stand here on whether it is true that all moral disagreement stems from religious dogma, irrationality, and so forth. My point is that disagreement among theorists over mathematical axioms raises doubts about the self-evidence etc. of mathematical propositions far more effectively than disagreement which can be easily explained away in terms of these features.[38]

2.9 Philosophy Everywhere

I have argued that our mathematical beliefs have no better claim to being a priori justified than our moral beliefs, realistically construed—contrary to

---

[37] Thanks to Joel David Hamkins for pressing me to make this contrast explicit.

[38] Leiter [2009] makes a similar point vis-à-vis disagreement among professional ethicists as compared to ethical disagreement among people more generally.

what is widely held. I have not argued that a priority is a useful notion, or that beliefs of either kind are a priori, in fact. I have argued that if there is a lack of parity between the justification of our moral and mathematical beliefs, it cannot stem from the sorts of features that are widely supposed to distinguish mathematics from morality, and philosophy more generally— such as that only mathematical propositions are self-evident or provable. One upshot of the discussion is that the extent of disagreement in an area, in any ordinary sense, may have little epistemic consequence—contrary to what is widely supposed.

All manner of established sciences are up to their ears in controversial philosophy, by which I just mean explicit theories about their subject matter. Moreover, it is hard to see why agreement among those who are ignorant of such theories should count for much. If so, then, *for epistemological purposes*, all such sciences may be comparably controversial, since the philosophy they bottom out is.[39] Despite the conventional wisdom that philosophy is controversial while the sciences are not, it might be more accurate to say that all areas of inquiry are controversial, in the sense that should be of interest to epistemologists, because an explicit formulation of an area's findings will generally be up to its ears in controversial philosophy.[40]

Consider physics. If (pure) mathematics is the queen of the sciences, then physics is not far behind. But physics remains notoriously controversial among experts. Of course, we all know that physics is not "finished" and may never be. But even fundamental theories, such as quantum mechanics, remain highly controversial *under a fixed interpretation*. Sean Carroll recalls,

> At a workshop attended by expert researchers in quantum mechanics....
> Max Tegmark took an…unscientific poll of the participants' favored
> interpretation of quantum mechanics....The Copenhagen interpretation
> came in first with thirteen votes, while the many-worlds interpretation

[39] Contra McGrath [2007, 95].

[40] Frances [2005] uses a related observation to give an empirical argument for skepticism. (Note that, if this is right, more nuanced norms of deference in the sciences may be called for. Many would deny that we ought to defer to the majority of philosophers on matters of epistemology. But the problem is not that philosophers fail to be more knowledgeable of the issues or more meticulous in their thinking about them than the rest of us. The problem is that there seems to be no principled way to decide *to whom to defer*. Epistemology is notoriously controversial. We should not, it would seem, defer to the majority of philosophers who happen to favor epistemological internalism, when, among those *specializing* in the internalism/externalism debate, a disagreement rages, with no apparent method to resolve it. In light of the above, the same may be true of at least some of the sciences. Even if the majority of mathematicians accept the Axiom of Choice, this is not good reason for us to accept it if those specializing on Choice continue to disagree, and lack a method to resolve their dispute.)

came in second with eight. Another nine votes were scattered among other alternatives. Most interesting, eighteen votes were cast for "none of the above/undecided." And these are the *experts*.

[2010b, 402, n. 199, my emphasis]

The term "interpretation" is a bit of a misnomer. Alternative interpretations of quantum mechanics amount to radically different conceptions of the physical world, not to subtly different spins on a common narrative. What is common among experts in quantum mechanics is reliance on a certain mathematical recipe for predicting the results of experiments—not a theory of the physical world. As Murray Gell-Mann puts it, "Quantum mechanics is not a theory, but rather a framework within which we believe any correct theory must fit" [quoted in Mulvey 1981, 170]. What the so-called wave function is, whether it is a component of physical reality, whether it collapses as a result of observation, and, if it does, how observation in the relevant sense relates to consciousness—these are among the fundamental questions that remain intensely disputed.

I mentioned that mathematicians are overwhelmingly focused on questions of logic, rather than on questions of non-logical truth. It might be wondered, then, whether the above points apply to logic itself. If not, then perhaps one should contrast morality with metalogic (the theory of what follows from what), not mathematics. But, on the contrary, metalogic seems to be no different from mathematics in pertinent respects. While most of us lack a systematic view on the question of what follows from what, those who are familiar with arguments and evidence bearing on the matter disagree much like those working in the foundations of mathematics (indeed, there is much overlap among the researchers). Classical logicians accept the Law of the Excluded Middle, while intuitionists do not. Paraconsistent logicians reject what is common ground among classicists and intuitionists, and dialetheist logicians go further and accept the actual truth of non-meta-logical propositions rejected by the others. There are advocates of all manner of alternative views as well.[41]

And so it goes. Pertinent knowledge seems to make "something so simple as not to seem worth stating" into "something so paradoxical that no one will believe it"—whether that thing be a moral, mathematical, or even logical proposition. We would like to quarantine the arguments which generate the

---

[41] See Williamson [2012] for an argument that not even logic affords a "neutral" arena in which to assess arguments, philosophical or otherwise.

puzzles, since, as Hilbert says, "If mathematical thinking is defective, where are we to find truth and certitude" [1983/1936, 191]? But we can only do so in bad faith. That it strikes us as true that any two sets are of the same cardinality, or one is of greater cardinality than the other, *bracketing* "philosophical" arguments to the contrary should count for no more than that it strikes us as true that the sun orbits the earth, bracketing scientific ones to the contrary.

I conclude that our mathematical beliefs have no better claim to being (defeasibly) a priori justified than our moral beliefs. But there is a less common way to think about the justification of our mathematical beliefs. According to it, they are justified empirically. I turn to that now.

3

Observation and Indispensability

I have argued that our mathematical beliefs have no better claim to being (defeasibly) a priori justified than our moral beliefs, realistically construed. In particular, their contents have no better claim to being self-evident, provable, plausible, initially credible, or epistemically analytic than the contents of our moral beliefs. There does not even seem to be an *epistemologically important* sense in which their contents are less controversial than the contents of our moral beliefs, realistically construed. If our mathematical beliefs are justified in the way that they are widely supposed to be, then our mathematical beliefs have no better claim to being (defeasibly) justified than our moral beliefs, realistically construed. In fact, analogous considerations seem to show that they have no better claim to being justified than our philosophical beliefs generally.

There is a less familiar account of the justification of our mathematical beliefs, however. As Gottlob Frege [1980/1884, 26] emphasizes, mathematics is not an isolated science. It applies to the non-mathematical world. Some have tried to use this fact to argue that our mathematical beliefs are (perhaps additionally) *empirically justified*, thanks to the role that their contents play in our best empirical theories. Since the contents of our moral beliefs do not seem to play an analogous role, there may still be room to argue that our mathematical beliefs are (defeasibly) justified while our moral beliefs are not. Perhaps only the former are empirically justified.

In this chapter I argue that our mathematical beliefs have no better claim to being empirically justified than our moral beliefs, focusing on Gilbert Harman's influential argument to the contrary. I show that Harman's reasons to think that the contents of our moral beliefs fail to be implied by our best empirical scientific theories serve equally to show that the contents of our mathematical beliefs do too, realistically construed. I then formulate a better argument for a lack of parity between the cases, in terms of indispensability. I argue that while the "necessity" of mathematics is no bar to producing a mathematics-free alternative to our empirical scientific theories, contra a recent objection of Timothy Williamson, the contents of our arithmetic beliefs, realistically and even objectively construed, seem to be indispensable to

Morality and Mathematics. Justin Clarke-Doane, Oxford University Press (2020). © Justin Clarke-Doane.
DOI: 10.1093/oso/9780198823667.001.0001

every explanation whatever at the level of metalogic—the theory of what follows from what. But this still fails to show that those beliefs, much less the range of mathematical beliefs that we actually have, are empirically justified. Surprisingly, however, the range of moral beliefs that we actually have may be, albeit in a different way. Unlike mathematics, there is no apparent ground on which to rule out so-called "moral perceptions" as perceptions per se (at least in the sense that we can perceive any high-level properties). I conclude with the prospect that there is no principled distinction between intuition and perception and, hence, between a priori and a posteriori justification.

3.1   Indispensability

In an influential book, Gilbert Harman writes,

> In explaining the observations that support a physical theory, scientists typically appeal to mathematical principles. On the other hand, one never seems to need to appeal in this way to moral principles. Since an observation is evidence for what best explains it... there is indirect observational evidence for mathematics. There does not seem to be observational evidence... for basic moral principles.   [1977, 9–10]

By an "observation," Harman means an "immediate judgment made in response to the situation without any conscious reasoning" [1977, 208], where a judgment, in turn, is a mental event, not a propositional content. More exactly still, an observation, in the present sense, is the fact that such a mental event occurred. So, our observation that there is a piece of paper in front of us is the fact that we spontaneously judged that there was a piece of paper there (at time, t) without any conscious reasoning. It is not, in other words, the fact that there is a piece of paper in front of us. Assuming that some moral truths can explain others, it would be unsurprising if such and such moral principles explained the *contents* of our spontaneous moral judgments.

Harman is naturally read as sketching an argument that our mathematical beliefs are empirically justified, but our moral beliefs are not. If Harman's reasoning has relevance to the realism–anti-realism debate, then he intended a stronger conclusion. The stronger conclusion is that our mathematical beliefs, *realistically construed*, are empirically justified, but our moral beliefs, so construed, are not. By "realistically construed," I mean construed in accord with the schemas of F-Aptness, F-Truth, F-Belief, F-Independence, and F-Face-Value from Section 1.2.

Let us call the argument above *Harman's Argument*. Understood in this way, Harman's Argument depends on three premises (the rationale for the labels will become clear below):

Quine–Putnam Thesis: The contents of at least some of our mathematical beliefs, realistically construed, are implied by the best explanation of at least some of our observations.

Harman's Thesis: It is not the case that the contents of any of our moral beliefs, realistically construed, are so implied.

Quine–Harman Thesis: Our belief is empirically justified just in case its content is implied by the best explanation of some of our observations.[1]

It follows from these three premises that at least some of our mathematical beliefs are empirically justified, but none of our moral beliefs is, realistically construed.

3.2 The Quine–Putnam Thesis

In a memorable passage from "Two Dogmas of Empiricism," W. V. O. Quine writes,

> Objects at the atomic level and beyond are posited to make the laws of macroscopic objects, and ultimately the laws of experience, simpler.... Moreover, the abstract entities which are the substance of mathematics... are another posit in the same spirit. Epistemologically these are myths on the same footing with physical objects...neither better nor worse except for differences in the degree to which they expedite our dealings with sense experiences.   [1951b, 42]

Quine is sketching an empiricist epistemology of mathematics. Unlike traditional empiricist epistemologies of mathematics, Quine's does not just

---

[1] Note that all of these theses could be restated in the language of truth by uncontroversial instances of the T-schema. For instance, an equivalent formulation of Harman's Thesis is that our belief is empirically justified just in case its truth is implied by the best explanation of some of our observations.

promise to explain how our belief in trivialities like that 2 + 2 = 4 could be justified. It promises to explain how our belief in standard claims of modern mathematics, like the Axiom of Infinity or the Mean Value Theorem, could be. It is not that we directly observe that, for example, there is an inductive set. Nor, contra Mill [2009/1882], is the hypothesis that there is such a set justified by means of an enumerative induction, as our belief that all ravens are black is.[2] Our mathematical beliefs are justified in the way that our beliefs about theoretical physical postulates, like electrons or gravitational waves, seem to be justified. They "are to be vindicated…by the indirect systematic contribution which they make to the organizing of empirical data in the natural sciences" [1958, 4]. That is, they are justified by an *inference to the best explanation* of our observations. Hartry Field writes,

> [T]he theories that we use in explaining…facts about the physical world not only involve a commitment to electrons and neutrinos, they involve a commitment to numbers and functions and the like. (For instance, they say things like 'there is a bilinear differentiable function, the electromagnetic field function, that assigns a number to each triple consisting of a space-time point and two vectors located at that point, and it obeys Maxwell's equations, and the Lorentz force law.')…There seems to be no possibility of accepting electrons on the basis of inference to the best explanation, but not accepting mathematical entities on that basis.…
>
> [1989, 16–17]

Field's point is that standard formulations of our best empirical scientific theories, from physics to economics, are up to their ears in mathematical language. Interpreted at face value, these theories refer to numbers, sets, functions, and so on. To be sure, they rarely directly appeal to the axioms of mathematics. But they still presuppose them, if only implicitly. For example, a physical theory that quantifies over real numbers presupposes the axioms of real analysis, since those axioms govern the numbers over which the theory quantifies. Moreover, since the best account of the real numbers is given by set theory, a physical theory that quantifies over real numbers presupposes some axioms of set theory too.

Of course, our best empirical scientific theories are not primarily focused on explaining mental events. Mechanics is focused on physical objects, not

---

[2] See Kitcher [1985] for a much more sophisticated epistemology of mathematics inspired by Mill.

on our observations of them. But, as Harman and Quine emphasize in the passages above, those theories have at least an indirect role in explaining observations. This is why it makes sense to call them empirical theories. So, if we grant that our best empirical scientific theories together constitute (or approximate) the best explanation of our observations, then, by the Quine–Harman Thesis, we can infer that the contents of some of our mathematical beliefs, realistically construed, are empirically justified.

3.3 Harman's Thesis

Harman's Thesis says that no analogous argument is possible in the moral case. Harman writes,

[Y]ou do not seem to need to make any assumptions about any moral facts to explain the occurrence of the so-called moral observations. In the moral case, it would seem that you need only make assumptions about the psychology or moral sensibility of the person making the moral observation. [1977, 6]

[A]n assumption about moral facts would seem to be totally irrelevant to the explanation of your making the moral judgment that you make. [1977, 7]

Sayre-McCord summarizes the contrast to which Harman gestures as follows:

Just as mathematics is justified by its role in explaining...observations, moral theory might similarly be justified by its role in explaining...observations. But the problem with moral theory is that moral facts seem not to help explain the making of *any of our observations.*

[1988, 443, emphasis in original][3]

(While Harman focuses on whether propositions of a kind, F, explain our *F-observations*, what matters for the Quine–Harman Thesis is whether they explain our observations of *any* kind.)

---

[3] Note that Sayre-McCord himself is critical of Harman's discussion.

The claim that "an assumption about moral facts would seem to be totally irrelevant to the explanation of your making the moral judgment that you make" is overblown. As Nicholas Sturgeon complains, "[i]t is commonplace to explain people's actions by appeal to moral states of character, good and bad, just as it is commonplace to hear social revolutions... attributed to the combined effects of poverty and injustice" [2006, 245].[4] But Harman does not just dismiss moral explanations out of hand. He sketches two arguments for thinking that they are never true. The first is that we have no idea how they could work, because we have no idea how moral properties could cause anything. After all, "[w]e cannot just make something up, saying, for example, that the wrongness of [an] act affects the quality of the light reflected into [our] eyes, causing [us] to react negatively" [Harman 1986, 63, italics in original]. The problem, then, according to Harman, is that "there does not seem to be any way in which the actual rightness or wrongness of a given situation can have any effect on your perceptual apparatus" [1977, 8].

The first thing to note about this argument is that it involves the confusion described in Section 1.5. It is no part of moral realism that there is literally an entity, wrongness (or The Wrong).[5] By that reasoning, we should ask about the actual rock-ness of a rock. How, we might query, could *the actual rock-ness* of the rock affect our perceptual apparatuses? The sensible view in the neighborhood is, of course, that it is *the rock*—or, more accurately still, the event of the rock's being there (at time t), or the fact that the rock was there—which affects people's perceptual apparatuses. Similarly, the sensible moral realist view is that it is *the wrong action*—or the event of the wrong action's occurring, or the fact that the wrong action occurred—which affects people's perceptual apparatus.[6] So, the question is not whether

---

[4] One of Sturgeon's most famous examples of a moral explanation concerns the rise of abolitionism. He writes, "An interesting historical question is why vigorous and reasonably widespread moral opposition to slavery arose for the first time in the eighteenth and nineteenth centuries, even though slavery was a very old institution.... There is a standard answer.... It is that chattel slavery... was much worse than previous forms of slavery..." [1984, 64–5].

[5] Again, the same holds for reasons.

[6] This confusion is commonly present in the work of non-naturalist moral realists too. Ronald Dworkin writes, "The idea of direct impact between moral properties and human beings supposes that the universe houses, among its numerous particles of energy and matter, some special particles—morons—whose energy and momentum establish fields that at once constitute the morality or immorality, or virtue or vice, of particular human acts and institutions and also interact in some way with human nervous systems so as to make people aware of the morality or immorality or of the virtue or vice" [1996, 104]. Similar allegations arise in the context of so-called Genealogical Debunking Arguments, to be discussed in Chapter 4. For example, Sharon Street argues that it would not have promoted reproductive fitness to have

"the actual wrongness" could cause anything. The question is whether, e.g., the event of the kids acting wrongly could.[7]

The question is controversial. Various theorists have argued that it could (Boyd [1988], Brink [1989, 189], Railton [1986, 173], Roberts [2016], Sturgeon [1984]). But what matters is that there would be an analogy with the mathematical case *if they were wrong*. Harman himself says,

> We do not even understand what it would be like to be in casual contact with the number 12, say. Relations among numbers cannot have any more effect on our perceptual apparatus than moral facts can.   [1977, 10]

Indeed, the view that (pure) mathematical facts would be causally inert is one of least controversial in philosophy. It is precisely mathematical facts' causal inertness that motivates work on fictionalist accounts of its application in empirical science. As Field writes,

> [E]*ven on the assumption that mathematical entities exist*, there is a *prima facie* oddity in thinking that they enter crucially into explanations of what is going on in the non-platonic realm of matter. . . . [T]he role of mathematical entities, in our explanations of the physical world, is very different from the role of physical entities in the same explanations . . . [because f]or the most part, the role of physical entities . . . is causal: they are assumed to be causal agents with a causal role in producing the phenomena to be explained.   [1989, 18–19, italics in original]

Field rejects the literal truth of explanations which imply atomic mathematical sentences, realistically construed, for much the reason that Harman rejects moral explanations. He writes,

> [I]t seems to me that . . . one wants to be able to explain the behavior of the physical system . . . without invoking . . . entities (whether mathematical or non-mathematical) whose properties are irrelevant to the behavior of the

true beliefs "about" the realist's moral values, given that "[o]ne of course cannot run into, or eat or be eaten by, reasons or values" [2016, 320].

[7] One might think that we could still ask about the property in virtue of which the rock has a given effect (e.g., its weight rather than its colors), and so can ask if that property is responsible for the action's effect on us. But in the context of Quinean nominalism about universals, such questions make no sense. See, again, Section 1.5.

system being explained). If one cannot do this, then it seems rather like magic that the…explanation works.   [1985, 193]

In fact, the argument from causal inertness is *more* plausible in the mathematical case. In the moral case, our atomic spontaneous judgments, such as that what such and such people are doing is wrong, at least *counterfactually co-vary* with the truths (Sturgeon [1984]). Had they not been doing something wrong, the world would have been different in non-moral respects, and our moral beliefs would have reflected the difference—since the closest worlds in which they are not doing something wrong are presumably still worlds in which the "explanatorily basic" moral truths, which fix the supervenience of the moral on the non-moral, are the same. Had the people not been doing something wrong, then they would have been, say, petting the cat instead of pouring gasoline on it, and we would no longer have judged that they were doing something wrong. On the other hand, had, for example, 2 not been prime, it seems that our judgments would have been unaffected (see Section 4.3 and, especially, 5.6). One might protest that the latter proposition is necessary, while the proposition that the people are doing something wrong is contingent. So, the counterfactual "had 2 not been prime, we would *not* have believed that it was" is true too— albeit vacuously. But, as we will see in Section 3.4, even if the mathematical truths are necessary in some sense, this is not a sense in which counterpossibles are vacuous.[8]

So, *if* the reason for believing Harman's Thesis is that it is hard to see how moral facts could cause anything, *then* we should reject the Quine–Putnam Thesis out of hand. What of Harman's second argument for that thesis? Harman is not an error theorist about all apparently causally inert properties. Some properties, like the property of being a chair or a restaurant, can be reduced to natural ones. When they can be, Harman suggests, truths ascribing them are implied by the best explanation of our observations, even though we do not appear to appeal to them in those explanations. Harman's second argument against moral explanations is that, in addition to appearing to be causally inert, moral properties are not reducible to natural ones.

---

[8] It might be thought that our judgment that the number of kids torturing the cat is equal to 2 co-varies with the facts. But this is an impure, not pure, mathematical judgment. (The pure/ impure distinction crosscuts the atomic/non-atomic one.) Moreover, the assumption about closeness of worlds underlying this thought is doubtful. See my [2019, Section 7].

Harman does not say what he means by "reduction" (or "natural"). He might take reducing an allegedly non-natural property, F, to natural ones to require explicitly defining "F" in terms of natural predicates. But, if that is what it takes, then it is doubtful that such everyday properties as being a restaurant are, in fact, reducible to natural ones (even if we allow infinite disjunction). So, perhaps Harman might conceive of property reduction on the model of Kripke [1980, Lecture III], according to which properties are regarded as entities with identity conditions. But that model only seems to make sense in the context of realism about universals. What would it mean to say that the entity, goodness, *really is*, say, the entity, pleasure maximization, assuming that there are not really any such entities? Presumably, it would have to mean something like "is good" *means* maximizes pleasure—that is, that "is good" is definable in terms of certain natural predicates. Again, however, the irreducibility of moral properties to natural ones in this sense is no objection to their "existence." Finally, Harman might take reducing moral properties to natural ones to involve showing that they are necessarily coextensive with natural properties, or at least that they supervene on natural properties. But, in that case, there is no apparent obstacle to reducing them (Brown [2011], Jackson [1998], Ridge [2007]). On the contrary, virtually everyone agrees that moral properties would supervene on natural properties.[9]

Although Harman's suggestion that moral properties cannot be reduced to natural ones is unmotivated, let us once again grant it. It still fails to establish a lack of parity with the mathematical case. Reductions of moral properties to natural ones have at least been *proposed*. For instance, Peter Railton suggests that an "individual's good consists in what he would want himself to want, or to pursue, were he to contemplate his present situation from a standpoint fully and vividly informed about himself and his circumstances,

---

[9] Rosen [Manuscript] rejects supervenience with respect to metaphysical possibility, but also calls it "the least controversial thesis in metaethics." It might be thought that the problem is that the moral realist cannot *explain* the supervenience. But this cannot be right if it means showing that moral predicates are definable in terms of natural ones, or showing that moral properties really are natural ones, or showing that they are necessarily coextensive with natural ones. If it does not mean this, then I am not sure what it means. Blackburn [1971] appears to suggest that in unproblematic cases of supervenience there is a conceptual entailment from a specification of the subvenient facts to a specification of the supervenient ones. But, even if that were so, it is hard to see how that could afford the needed explanation. We cannot explain the *metaphysical* fact of supervenience in terms of the *epistemological* fact that there is a connection between some concepts—lest we revert to the metaphysical concept of analyticity, which Quine [1951] discharged. (Perhaps one could appeal to the ideology of grounding or constitution? The fact that some Bob Evans is a restaurant is grounded in the fact that it is F, G, H ... (for uncontroversially natural predicates "F", "G", "H" ...). Again, just as a moral realist need not be a realist about universals, she need not be a realist about grounding.)

and entirely free of cognitive error or lapses of instrumental rationality" [2003/1993, 52].[10] By contrast, it is hard to even *formulate* an analogous proposal in the mathematical case. It has certainly been argued that natural numbers are reducible to cardinality properties of collections (Bealer [1982], Jubien [2006], Lowe [1993 and 1995]). But such reductions assume realism about universals. Again, unlike our moral theories, our mathematical theories primarily *quantify over* mathematical entities. They do not just predicate them (if they predicate them at all). More exactly, such reductions assume *Platonism* about universals—that is, that properties exist independent of their instantiation. The reason is not that numbers exist necessarily (more on this below), contra Shapiro [2000, 21–4]. The reason is that our mathematical theories imply the existence of cardinalities, such as κ from Section 2.3, which, qua properties, are surely not exemplified in nature.[11] But *Platonistically construed*, properties are widely agreed to exist "outside" spacetime—that is, to lack spatiotemporal properties—and to be causally inert (Bealer [1982], Jubien [1997, ch. 3], Wolterstorff [1970]) Indeed, if one could manage to explain how *Platonic properties* have spatiotemporal location, or participate in the causal order, then one could presumably explain how sui generis mathematical entities do too. If Platonic properties count as natural, then there seems to be no epistemological reason to *care* whether moral properties reduce to natural ones.

The upshot is that if the argument for Harman's Thesis is that we have no idea how moral properties could cause anything, or how they could be reduced to natural ones, then the argument works a fortiori to show that the Quine–Putnam Thesis is false. But Harman, and many of those working on moral explanation, tend to conflate the question of whether the contents of our moral beliefs are implied by the best explanation of our observations with that of whether moral properties are causally efficacious or natural.[12] As the mathematical example shows, these things may come apart. The puzzle in the mathematical case is precisely that the contents of our mathematical

[10] A naturalist might worry about the appeal to rationality. See Boyd [1988] and Jackson [1998] for purportedly naturalist views that avoid this.

[11] This point seems to be to be overlooked by so-called *physicalist* accounts of mathematics, such as Maddy [1990]. While (impure) *sets* on Maddy's view are concrete, Maddy identifies numbers, not with sets, but with *properties* of them. So, the same reasoning should compel her to regard numbers as Platonic entities, contra the spirit of the book. Even the view that sets are concrete is hard to sustain. See Balaguer [2016, § 5].

[12] See, for example, Majors [2007]. Sturgeon himself seems to have assumed that, if some facts explain observable phenomena, then they are natural, and that natural facts enter into the causal order. See Sturgeon [2006].

beliefs appear to be implied by the best explanation of our observations, realistically construed, *even though mathematical entities are neither causal nor natural in any useful sense.*[13]

So, it would be interesting if there were an argument for Harman's Thesis that did not depend on considerations of causation or naturalness. It is, after all, not beyond dispute that we should take the notion of causation seriously. Bertrand Russell underscored that "[t]he concept 'cause,' as it occurs in the works of most philosophers, is one which is apparently not used in any advanced science" [1948, 471]. Is there any argument for Harman's Thesis that avoids this concept?

There is one such argument. Moral facts are independently mysterious. John Mackie says that "they would be entities...of a very strange sort, utterly different from anything else in the universe" [Mackie 77]. Parsimony considerations thus favor non-moral explanations over moral ones, other things being equal. More explicitly, they favor non-moral explanations over moral explanations, given that there are any. And there certainly *are* non-moral explanations. It is not as if science would grind to a halt if we were precluded from appealing to moral explanations! On the other hand, there seem to be no alternatives to many mathematical explanations at all (Sober [1993]). So, even if mathematical facts are mysterious (whether in the same way or in another), this does not show that they fail to be implied by the best explanation of our observations. It may be that, in this case, mysterious facts are the only ones available to which to appeal. In a slogan, moral explanations are *dispensable*, while mathematical ones are not.

I will ultimately argue that there is something to this. But, first, let me discuss a way to resist.

3.4 Instrumentalism and Modality

There is what appears to be a simple way to generate a non-mathematical alternative, T*, to any mathematical explanation, T, of some empirical phenomenon. T* is not just empirically indistinguishable from T. It is *concretely* so—that is, it is indistinguishable from T with respect to its implications for the concrete reality. Consequently, it would seem to carry the same causal information, and, for all that has been said, may be comparably simple,

---

[13] See Kitcher [1985, 104–5] for related discussion.

explanatory, fecund, and so forth. Gideon Rosen outlines (but does not endorse) the strategy as follows:

> [S]uppose ours is a numberless world and that [any scientific explanation which implies mathematical sentences, realistically construed, is] therefore false. If we were concerned to speak the truth, we would never countenance its assertion. But the fact is, we are rarely concerned to speak the truth. Our unhedged assertoric utterances normally aspire to a weaker condition we call *nominalistic adequacy*. S is nominalistically adequate iff the [mereological sum of all the concrete objects that exist in the] actual world is an exact intrinsic duplicate of [the mereological sum of all the concrete objects that exist in] some world at which S is true – that is, just in case things are in all concrete respects *as if* S were true.
>
> [2001, 75, italics in original][14]

Let us call the narrator an *instrumental fictionalist*. Then it is tempting to dismiss instrumental fictionalism as being like scientific instrumentalism. When the scientific instrumentalist "accepts" a theory, T, which seems to speak of unobservable entities, like electrons, she only really believes that it is *in all observable respects as if* T were true. Similarly, when the instrumental fictionalist "accepts" a theory, T, which seems to speak of mathematical entities, she only really believes that it is in *all concrete respects as if* T were true. But instrumentalist fictionalism does not stand or fall with scientific instrumentalism unless unobservable physical entities stand or fall with mathematical ones. Prima facie, they do not. As Mary Leng puts it,

> Mathematical objects are…acausal and non-spatiotemporal…. These… features put them on a…different footing than electrons…[W]e should

[14] This strategy has been advocated by a variety of philosophers more or less explicitly. For instance, Leng writes, "A fictionalist about mathematics will not believe the literal truth of the (mathematically stated) utterances that are used to express our ordinary empirical theories, but will, instead, believe that those utterances get things right in their picture of how things are with their *non-mathematical objects* In short, we might say, a fictionalist will assume that our scientific theories are *nominalistically adequate*, but not that they are true" [2010, 180, emphasis in original]. Baker [2003], which criticizes the strategy, includes the following quotations: "Since sets are not supposed to be part of the world's spatio-temporal causal nexus, that nexus would be exactly as it is whether sets existed or not" [Horgan 1987, 281–2]. "Since mathematical objects are acausal, the existence or non-existence of mathematical objects makes no difference to the actual arrangement of concrete objects" [Cornwell 1992, 80]. "The basic reason for resisting abstract [objects]…is that the world we can know about would be the same whether or not they existed" [Ellis 1990, 328]. "If there were never any such things as [mathematical] objects, the physical world would be exactly as it is right now"[Balaguer 1999, 113].

expect that the observed phenomena would be very different on the hypothesis that there are no such things [as electrons]...But if such counterfactual considerations have force against those sceptical about the unobservable physical objects posited by our theories, no analogous counterfactual is available against those sceptical about the mathematical objects our theories posit. A mathematical realist who starts a challenge, "If there *were* no numbers, then..." will find it difficult to finish this supposed counterfactual in a way that could trouble those sceptical of mathematical objects.   [2010, 202, italics in original]

There is a worry for instrumentalist fictionalism, however. Timothy Williamson formulates it as follows.

The [instrumentalist fictionalist] reasons in effect about *how things would be if the mathematical theory were to obtain and concrete reality were just as it actually is.* Thus the conclusion corresponds to this counterfactual

(15) (M & A) []→ C

Here M is the mathematical theory [realistically construed], A says that concrete reality is just as it actually is, and C says something purely about concrete reality. Thus, the truth of the counterfactual seems to guarantee the truth of its consequent, even though its antecedent is false (by [instrumental fictionalist] lights), because the relevant counterfactual worlds are the same as the actual world with respect to concrete reality, which C is purely about. The trouble is that the [instrumental fictionalist] may well regard platonism as not just *false* but *metaphysically impossible*: for instance, the structure of the hierarchy of pure sets (if any) seems to be a metaphysically non-contingent matter.... [Instrumentalist f]ictionalists who implement their strategy with counterfactuals and regard the rival metaphysical theory as a useful but impossible fiction have therefore been compelled to deny orthodoxy about counterpossibles. (for instance, Dorr 2008)   [2017, 199–200, italics in original]

Williamson argues that the instrumental fictionalist must agree that we can reliably reason about the concrete world using standard formulations of our best empirical scientific theories, even if these are literally false. Hence, she must believe in the truth of counterfactuals conditionalizing on the false mathematics.[15] However, (pure) mathematical propositions are supposed to have whatever truth-values they have of necessity. Moreover, on the standard semantics for counterfactuals, *had it been the case that P, it would have been the case that Q*, written (P []—> Q) is true if the "closest" or "most similar" possible worlds in which P is true are worlds in which Q is true too. Consequently, if P is a false mathematical proposition, and Q is any proposition whatever, (P []—> Q) is (vacuously) true. The whole question of what the world would be like had our standard mathematical theories been true, therefore, gets trivialized.

Such arguments trade on an ambiguity in the word "necessary." Williamson notes that the standard account of counterfactuals is plausible only so long as "necessity" is taken to mean "the maximal objective" notion of necessity [2016, 460]—where a maximal objective notion of necessity, [M], is such that, for any other objective notion of necessity, [N], and for any proposition, P, [M]P → [N]P, but not conversely. It is not actually immediate that there *is* a maximal objective notion of necessity (Clarke-Doane [2019, § 8]), but let us suppose that there is. What does "objective" mean? An objective notion of necessity "is what the modal words express when they are *not* used in any epistemic or deontic sense..."[Strohminger and Yli-Vakkuri 2017, 825, emphasis in original]. For instance, *it is known that* and *it is required that* are not objective necessity operators. But this still leaves tremendous leeway. For instance, (first-order) logical possibility is not defined in terms of epistemic or deontic concepts, but virtually every proposition of traditional metaphysical interest is logically contingent.[16] Williamson adds that objective notions of necessity are also "not sensitive to the guises under which the objects, properties, relations and states of affairs at issue are presented" [2016, 454]. This means that "identity [and distinctness are] simply objectively necessary..." [2016, 454]. Although there are reasons to resist (Clarke-Doane [Forthcoming]), let us grant this requirement as well. Still, *no existentially quantified mathematical propositions are necessary relative to a maximal objective notion of necessity* (even assuming their actual truth). For instance, they are not necessary relative to a notion of logical necessity corresponding to a quantified modal logic which fails to imply that everything which actually exists necessarily does. An example would be the one

---

[15] I do not actually believe that the "hence" is warranted. But I will not argue the point here.
[16] Even identities are contingent in the context of S5 if, e.g., one indexes the identity rules in the proof system to worlds as one does the other logical rules. See [Girle [2009, ch. 7]].

obtained from the original variable domain quantified modal logic of Kripke [1963]. Although these validate true identities, they validate virtually no other sentences of metaphysical interest.[17]

I am not suggesting, with Quine [1953], that we lack a coherent notion of metaphysical necessity, according to which such things as the mathematical truths are necessary. We could even define such a notion using a maximal objective notion of necessity, [M] (assuming, with Williamson, that there is such a notion). We could simply tack the metaphysically necessary truths on as "modal axioms" (Sider [2011, ch. 12]). Using [M], we could then define metaphysical necessity, [M*], by saying that $[M*]P \leftrightarrow [M](T \to P)$, where T is the (perhaps infinite) conjunction of propositions we want to call metaphysically necessary. Alternatively, there may be a less gerrymandered set, T, such as generalizations of the "essential truths" (Fine [1994], Hale [2013], Kment [2014]), from which the mathematical truths follow. Whether we use the term "metaphysical necessity" for a maximal objective notion of necessity or for this restricted notion is immaterial. What is important is that, even by Williamson's lights, counterpossibles are only vacuous when their antecedents are maximally impossible. Since mathematical truths are not maximally necessary, the argument that any counterfactual conditionalizing on a false mathematical proposition must be vacuous is simply fallacious.[18]

3.5 Indispensable Mathematics and Metalogic

Although the above objection to instrumental fictionalism fails, there *are* two problems with the view. The first is that we lack a clear conception of the *concrete world*, bracketing the mathematical predicates true of it. For example, Field and Leng suggest that electrons are concrete, while mathematical entities are not. But, to borrow an example from Putnam [2012, 196], what would it even mean to say that it is *as if* "the number of electrons in the box is indeterminate, but the state is $1/\sqrt{2}$(two electrons in the box) + $1/\sqrt{2}$(three electrons in the box)" *in all concrete respects*? By Born's Rule, the probability of finding two (three) electrons upon looking in the box is ½. Let us grant, what is certainly questionablec, that there is no problem understanding what the world would be like in concrete respects if the

---

[17] Actually, the sense in which they validate even true identities is vexed. See Varzi [Manuscript].
[18] For more on the modality, see the Section 7.2 of the Conclusion.

world is probabilistic in this way. It remains to characterize the box when unexamined. Its state then is a superposition between having different numbers of electrons in the box. As standardly conceived, this amounts to the mathematical fact that the state vector of the system is not an eigenvector of the operator representing the property of having a certain number of electrons in the box. It is far from clear that we will ever have an attractive non-mathematical characterization of superpositions. While non-mathematical alternatives to classical and relativistic gravitation theory do exist (Field [1980], Burgess and Rosen [1997, Part II], Arntzenius and Dorr [2012]), however objectionable they may be, quantum mechanics remains recalcitrant.[19] One might have hoped that the instrumental factionalist could have avoided commitment to its nominalistic formulation.[20]

It is tempting to rejoin that the fact that we do not know what the world would be like in concrete respects if a sentence like the above were true just shows that we do not yet know *what quantum mechanics says*. We characterize superpositions in mathematical terms because we do not know what they really are. And, indeed, on some interpretations of quantum mechanics, such as Everett's interpretation, it may be possible to factor out the "concrete content" of the theory. But we cannot just assume that quantum mechanics says something that can be factored into concrete and mathematical components in this way. The world may be stranger than that![21]

The second problem with instrumental fictionalism is that some mathematical truths seem to be indispensable to *metalogic*—the theory of what follows from what.[22] But we seem to be committed to metalogical truths in virtue of being committed to any theories whatever—*even the non-mathematical surrogates to which fictionalists appeal*. For any theory, T, whether about mathematical entities or not, if we rationally believe T, then we must at least believe that T is *consistent*. That is, we must believe that a contradiction

---

[19] See Malament [1982] for related discussion. I say "attractive" because Craig's Theorem ensures that, for any first-order theory incorporating mathematical language, one can always cook up a recursively axiomatized non-mathematical theory with the same non-mathematical consequences. But such a theory generally has no theoretical appeal. See Craig [1953], as well as Putnam [1965] and Field [1980, 8], for discussion.

[20] See, however, Chang [2018] for some recent work toward a "non-mathematical" quantum mechanics.

[21] One approach to quantum mechanics rejects the search for an interpretation of the formalism altogether, regarding the search for a translation of it into natural language misconceived. As Freeman Dyson writes, "the important thing about quantum mechanics is the equations, the mathematics. If you want to understand quantum mechanics, just do the math. All the words that are spun around it don't mean very much" [quoted in Roychoudhuri 2007]. See Siegel [2018] for a recent polemic to this effect.

[22] See Putnam [1979/1994, 501] for an objection along the following lines.

does *not* follow from T. But, while the theory of what follows from what officially concerns sentences—or, more exactly, strings of symbols—the theory of strings is bi-interpretable with the theory of natural numbers. Hence, there is a way to "say" anything that we want to say about strings of symbols in the language of arithmetic. It might be thought that linguistic items like strings of symbols are nevertheless more epistemically innocent than numbers, realistically construed. But this would be a mistake. Strings are not concrete. For example, for every string there is a longer one, but there is no guarantee that for every concrete string there is a longer one. Also, the symbols out of which strings are made cannot literally be anything like the concrete items that we use to represent them. A concrete symbol has shape and extension. For instance, the *token*, "0," is oval in shape. But the *type* "0" cannot literally be oval in shape, because, as normally conceived at least, types have no spatiotemporal properties at all—just like numbers. The notion of strings also brings to mind geometrical intuitions which are out of place. A string is a sequence of symbols from the alphabet, such as 00001001. But, again, the first "0" cannot really be to the *left* of the first "1."

So, metalogic commits us to the theory of the natural numbers, or something just like it. But, actually, it commits us to more. As far as indispensability for physics goes, mathematics could be wildly non-objective (in the sense of Section 1.6). The case of geometry makes clear that, just because physics uses a mathematical theory of a kind does not mean that an alternative theory of the same kind must be false. When physicists embraced general relativity, and with it Riemannian geometry, they did not declare Euclidean geometry false, as a pure mathematical theory (Leng [2010], 81]). But now consider the theory of what follows from what. As I mentioned in Section 2.2, it is consistent to say false things about consistency. For example, if Peano Arithmetic (PA) is consistent, then so is PA + ~Con(PA), where "~Con(PA)" codes the claim that a contradiction follows from PA. A model of PA + ~Con(PA) is a model in which there is an infinitely long "proof" of a contradiction from PA. I put "proof" in quotes, because a proof must be finite. The model is wrong about finiteness.[23] Or that is what we would like to say. But if we hold that PA + Con(PA) and PA + ~Con(PA) are equally true of their intended subjects, like, say, (pure) geometry with the Parallel Postulate and geometry with its negation, then there will be no objective fact as to what counts as finite and, hence, no objective fact as to what counts as a proof in PA. Consequently, there will be no objective fact

---

[23] The "proof" has the length of a non-standard number (i.e., a number greater than all of 1, 2, 3 ...), which the model thinks is a natural number.

as to whether PA, or any theory which interprets it, including a regimented physical theory, is consistent! I do not mean that it might count as consistent if based on classical logic, and inconsistent if based on a non-classical alternative. I mean that there would be no objective fact as to whether it is *classically consistent*. Assuming that there *is* such a fact, some arithmetic truths, or truths just like them (about strings), *objectively construed*, are indispensable to virtually all theorizing.[24]

This argument does assume that we must construe the indispensable arithmetic claims (or string claims) realistically. However, unless we are willing to be expressivists, fictionalists, or constructivists about consistency facts, we will not reject Arithmetic-Aptness, Arithmetic-Truth, Arithmetic-Belief, or Arithmetic-Independence (in the sense of Section 1.2). Fictionalists tend to reject Arithmetic-Face-Value by taking consistency facts as primitive and construing "it is consistent that," not as a predicate of sentences, but as an operator, like negation (Field [1989, Introduction], Leng [2007]). And this does let us avoid reference to linguistic types, in addition to numbers. We can now write <L>P to mean *that it is consistent that P*, where this is not a sentence *about* "P," any more than "it is not the case that grass is green" is about the sentence "grass is green." But nothing is gained by this approach. While it is ontologically more parsimonious than mathematical realism, it just trades ontology for ideology in the form of primitive modal operators (Shapiro [1995]). The chief benefit of the trade is supposed to be epistemological (Field [1989, Introduction], Leng [2007]). It is supposed to avoid the problem of having to explain our epistemic access to "remote entities" like linguistic types. But, as I suggested in Section 1.5 and will argue in Section 5.2, this is confused (Clarke-Doane [2016b, § 2.2]). The problem is *not* the one anticipated in Field [1989, Introduction] and Leng [2007], that the explication of the alleged ideological primitives will still somehow make reference to abstract objects, so the apparent loss of ontology is illusory. The problem is that whatever epistemological puzzles arise for mathematical realism arise equally for the varieties of anti-realism which satisfy Mathematical-Aptness, Mathematical-Truth, Mathematical-Belief, and Mathematical-Independence, *holding fixed the amount of objectivity that the views postulate*.[25] Indeed, if this were not the case, then a moral

[24] The "objectively construed" arithmetic truths cannot be recursively enumerable, much less recursive. See Section 6.2.

[25] If an anti-realist view satisfying F-Aptness, F-truth, F-Belief, and F-Independence attributes *less* objectivity than a given realist view, then it may indeed be better placed to explain our epistemic access to mathematical facts. But, as we will see in Sections 5.9 and 6.1, the same is true of realist views that attribute less objectivity than an anti-realist view of this sort. In a

realist could avoid any such epistemological puzzle by simply being a nominalist about universals![26] Ontology is *epistemically* irrelevant.

There is, then, a case to be made that the contents of our moral beliefs, realistically construed, are dispensable to the best explanation of our observations in a way that the contents of at least some of our mathematical beliefs are not. Indeed, the contents of at least a fragment of our arithmetic beliefs appear to be indispensable, *objectively construed*. Let us simply grant, then, that the Quine–Putnam Thesis and Harman's Thesis are both true. It only follows that some of our mathematical beliefs are empirically justified, while none of our moral beliefs is, realistically construed, if the Quine–Harman Thesis is true. In fact, however, it fails in both directions.

3.6 Perception

In the quotation that began this chapter, Harman concludes:

> Since an observation is evidence for what best explains it…there is indirect observational evidence for mathematics. There does not seem to be observational evidence…for basic moral principles.

This suggests one direction of the Quine–Harman Thesis—namely, that our belief that P is empirically justified *only if* P is implied by the best explanation of some of our observations (where, again, our observations are facts that certain mental events occurred). However, a moment's reflection reveals that this cannot be right. Whatever empirical justification amounts to, one's belief that P had better be (defeasibly) empirically justified if one *sensorily perceives* that P, and one has no defeaters. Actually, it had better suffice that one *quasi-perceives* that P, and has no defeaters—where a quasi-perception that P is just like a perception that P, but quasi-perception is not factive (I use "perception" rather than "observation" so as to avoid confusion with Harman's notion). One can quasi-perceive that P, even if ~P. If we are, sadly, brains in vats, then our belief that there is a piece of paper in front

---

slogan: what matters *epistemically* is objectivity, not ontology—just as we will see in Section 7.2, what matters *methodologically* is objectivity, not ontology.

[26] See, again, Section 1.5.

of us is still (defeasibly) empirically justified. (It is not as if it would be a priori justified, just because there is no paper.)[27] If we were to get *evidence* that we are brains in vats, then this justification might be defeated. But empirical evidence can be misleading. Misleading quasi-perceptions afford paradigm cases.

(I am not suggesting that *the fact that one quasi-perceives that P* is empirical evidence for P. I am suggesting that *in* quasi-perceiving that P, one is *thereby* (defeasibly) empirically justified in believing that P, whether or not one even has the concept of quasi-perception, or perception. In the jargon, I am advocating a *dogmatist* account of empirical justification, in the tradition of Pryor [2000] and Huemer [2005, Pt II].)

Of course, even if the *only if* direction of the Quine–Harman Thesis is false, it might still be that we cannot quasi-perceive moral contents. If that were so, then, assuming the Quine–Putnam Thesis, and the *if* direction of the Quine–Harman Thesis (something that I will challenge shortly), our mathematical beliefs might still have better claim to being empirically justified than our moral beliefs. But what is a principled account of quasi-perception according to which moral quasi-perceptions could be ruled out? Suppose, for instance, that X quasi-perceives that P in situation, S, just when X observes it in Harman's sense—that is, just when X "immediate[ly] judg[es that P] in response to [S] without any conscious reasoning" [Harman 1977, 208]. If so, then we can quasi-perceive everything from the Cardinality Comparability Principle (CCP) to the fact that what some people are doing to a cat is wrong. Harman himself notes that "[i]f you round a corner and see a group of young hoodlums pour gasoline on a cat and ignite it, you do not need to conclude that what they are doing is wrong; you do not need to figure anything out; you can see that it is wrong" [1977, 4]. Harman promptly adds, "but you do not seem to need to make assumptions about any moral facts to explain the occurrence of…so-called moral observations" [1977, 6]. However, if, in order to quasi-perceive that P it had to be the case that P is implied by some explanation of your spontaneously judging that P, then, given that explanation is factive, quasi-perception would be factive. Perhaps *knowledge* that any explanation of your spontaneously judging that P fails to imply that P would *undermine* that belief, realistically

---

[27] Virtually any case of perceptually justified false belief would serve the purpose. See Section 3.9 for another. (Although this view seems to me hard to resist, it is denied by so-called "disjunctivists." See McDowell [2008] and Martin [1997].)

construed (Joyce [2008], Street [2006]). I will discuss this possibility at length in Chapter 4. But everyone should agree that quasi-perceptions, including perceptions per se, can be undermined.

Nor could one rule out moral quasi-perceptions by adding that our spontaneous judgments are not fully explained by our "competence" with the concepts. Our spontaneous judgment in Harman's example of people pouring gasoline on a cat is surely not so explained. The same even seems to be true of interesting mathematical propositions. Again, it is a dubious article of faith that all mathematical intuitions, including, for example, that CCP is true, stem from conceptual competence. Once we move beyond banalities like that $2 + 2 = 4$, it is not plausible that our concepts afford sufficient data points from which to "induct" the axioms of set theory.[28]

On the other hand, our spontaneous judgment that CCP is true does not seem to be based on the deliverances of our five senses. Even if locked off from an outside world, if we considered whether CPP is true, perhaps we would spontaneously judge that it was. Similarly, while paradigmatic sensory judgments are about our concrete surroundings, spontaneous mathematical judgments, realistically construed, concern the likes of numbers, sets, and functions. Perhaps, then, quasi-perceptions must involve the deliverances of the five senses, and must concern concrete, rather than abstract, things. Indeed, if there is a standard view of the matter, this is it (Bengson [2015], Bonjour [1997], Chudnoff [2013, Introduction], Lewis [1986, 108–15]).

But while this would suffice to rule out spontaneous mathematical judgments, realistically construed, as quasi-perceptions, it would do nothing to rule out all spontaneous moral judgments, like the one to which Harman alludes. To suppose otherwise would be to revert to the mistake about properties discussed in Section 1.5 that I have belabored. Again, moral sentences such as "those people are doing something wrong" are not *about* moral properties in the way that mathematical sentences like "2 is prime" are about mathematical entities. Unlike the latter, they are about entities whose existence is not presently in dispute, and which we certainly do perceive with our five senses—namely, people. If there is a sense in which they are "about" moral properties, it is just the sense in which every prediction whatever is about an abstract object.[29]

---

[28] See, again, Chapter 2, especially Section 2.7.

[29] Assuming Platonism about universals. See Balaguer [2016, 4.2–4.4] for discussion. Even assuming Aristotelian realism or trope theory, however, it would still turn out that *no* predictions are merely about ordinary concrete things under the suggested use of "about."

Albert Casullo suggests that quasi-perception might be a natural kind whose underlying nature can be determined by empirical investigation.[30] But, as Jeshion [2011] notes, if quasi-perception were a natural kind, then we would not count differently built creatures with phenomenologically similar inner lives as having quasi-perceptions. Instead, we would count whatever creatures had states with the same underlying nature as having these. But, on the contrary, we would do the opposite (Kripke [1980, 147]). And yet, even if Casullo were right, it is hard to see how this could be a reason to doubt that there are moral quasi-perceptions in particular.

Of course, some philosophers deny that, strictly speaking, we can even perceive that some people are, e.g., petting a cat. They deny that we perceive "high-level properties" (Siegal [2016, § 4.3]). But not even they would deny that what we do perceive in such situations renders our spontaneous judgment that some people are petting a cat (defeasibly) empirically justified (even if not immediately[31]). What matters is whether there is a principled difference between such a judgment and the one that the people are doing something wrong. And while it *could* be that the latter is empirically justified only if belief in some background moral principle is a priori justified, it is hard to think of an argument for this that would not work equally in the non-moral case. Consider a statement of the conditions under which someone pets a cat. Such a statement would not tell us that any cat has been petted. It would tell us *what it is* for someone to pet a cat. On what ground could one argue that belief in such a statement may be justified empirically, but belief in a statement of the conditions under which someone acts wrongly could never be? Even given that there fails to be a perceptual phenomenology associated with spontaneous moral judgments, that moral categorization does not correspond to an activity in perceptual parts of the brain, that spontaneous moral judgments are too slow to be perceptions per se, and that such judgments lack sufficient independence from background beliefs to be perceptual judgments, it would *still* not follow that our spontaneous moral judgments fail to be (defeasibly) empirically justified.[32]

---

[30] Casullo [2002 and 2011] officially concerns "experience," in the sense of the a priori/a posteriori distinction.

[31] See McGrath [2018] for the view that while we do not immediately perceive such things as that some people are petting a cat, what we do perceive immediately justifies our belief that this is so.

[32] Thanks to John Morrison for helpful discussion of these points.

Is there any other way to argue that "spontaneous moral judgments" fail to count as empirically justified (if one's spontaneous judgments that some people are petting a cat does)? One could always gerrymander the definition of "empirical justification" so as to exclude them. But, then, even if our spontaneous moral judgments failed to count as empirically justified, they might be like paradigmatic empirical judgments in epistemically important respects. Better to mark this directly by categorizing them with judgments like "those people are petting a cat".

3.7 Justification and Explanation

So, the "only if" direction of the Quine–Harman Thesis fails, and any apparent account of quasi-perception would seem to allow that our moral beliefs can be (defeasibly) empirically justified, even if their contents fail to be implied by any explanation of our observations. But if the "only if" direction of the Quine–Harman Thesis fails, the "if" direction—that is, the claim that if P is implied by the best explanation of our observations, then our belief that P is empirically justified—fails a fortiori. Even granted that all (true) explanations of our observations include mathematical propositions, realistically construed, such propositions tend to play a transparently *representational role* in empirical scientific explanations. Typically, they index, or stand for, the physical quantities that are the real focus of the theory.[33] As Joseph Melia writes,

> [When] we come to explain [physical phenomenon] F our best theory may offer as an explanation, 'F occurs because P is root(2) metres long'. But…though the number root(2) is cited in our explanation, it is the *length* of P that is responsible for F, not the fact that the length is picked out by a real number.   [2002, 76]

It would be *very surprising* if the representational apparatus of an empirical theory were empirically justified in the way that the rest of the theory is. It would leave it mysterious why empirical scientists do not seem to think twice about "postulating" the likes of functions, while the postulation of new physical entities is met with empirical scrutiny. Penelope Maddy notes

---

[33] "Typically" because, in certain cases, like that of quantum mechanics, this is less clear.

that "physicists seem happy to use any mathematics that is convenient and effective, without concern for the mathematical existence assumptions involved..." [1997, 155]. Of course, as Quine [1951b, § VI] argued, the representational apparatus of a theory makes substantive claims on the world too. But one can grant his point while denying that those claims are empirically justified along with the rest of the theory. Hilary Putnam conceded as much. He writes,

> My argument was never intended to be an "epistemology of mathematics." If anything, it is a *constraint* on epistemologies of mathematics from a scientific realist standpoint.   [2015, 63, emphasis in original]

The so-called Quine–Putnam Indispensability Argument was put forward by Putnam as a *dialectical stance*. His point was that one can be a scientific realist, or mathematical anti-realist, but not both. If one wants to be a scientific realist, in the strong sense of accepting the truth of our best empirical scientific theories, under a face-value interpretation, then one cannot also be a mathematical anti-realist, on pain of incoherence. One's mathematical anti-realism will infect one's scientific realism, either because one will be forced to reject as untrue canonical statements of scientific laws, or because one will be forced to regard them as mind-and-language dependently true (because the mathematics that they incorporate is). Indeed, Field takes Putnam's dialectical point to generate "the only non-question-begging argument *for* the view that mathematics consists of *truths*" ([2016, 2, italics in original]). And perhaps the real import of Harman's Thesis is better understood as being analogous. One *cannot* canvass a non-question-begging argument "*for* the view that [morality] consists of *truths*." Harman writes,

> Observation plays a part in science it does not appear to play in ethics, because scientific principles can be *justified*...by their role in explaining observations....   [1977, 10, emphasis added]

This would constitute a disanalogy between mathematics and morality, if it were true. But the disanalogy would be dialectical. There would be, in that case, a way to convince a very thoroughgoing kind of scientific realist to be a mathematical realist that does not have an analog in the moral case.[34] But this

---

[34] See, again, the Introduction.

dialectical point does nothing, by itself, to show that our mathematical beliefs have better claim to being empirically justified, realistically construed.

Note that this argument is consistent with the Quine–Putnam Thesis. It does not threaten the earlier considerations suggesting that we lack a determinate notion of the concrete world, or that science commits us to some mathematical truths, realistically construed. It merely shows that the Quine–Putnam Thesis does not secure the *empirical* justification of our mathematical beliefs.

Recently it has been suggested, in effect, that the Quine–Harman Thesis could be replaced by a more believable thesis. An analog to the Quine–Putnam Thesis then generates what has become known as the *enhanced* indispensability argument for mathematical realism (see Baker [2005] and Lyon and Colyvan [2008] for examples). The amended theses are:

Quine–Putnam Thesis*: The contents of at least some of our mathematical beliefs, realistically construed, play an *explanatory role* in the best explanation of some of our observations.

Quine–Harman Thesis*: Our belief that P is empirically justified if and only if P plays an *explanatory role* in the best explanation of some of our observations.

The "only if" direction of the Quine–Harman Thesis* fails for the reason that the only if direction of the Quine–Harman Thesis fails. We may quasi-perceive that P even if P is not implied by the best explanation of any of our observations, whether in an explanatory way or not. But, while still controversial (Leng [2010, § 9.3 and 9.4]), the "if" direction of the Quine–Harman Thesis* at least now has some credibility. When exactly does P play an explanatory role in an explanation? There is no settled answer (Saatsi [2016]). But there are paradigm cases. One of the most referenced concerns the life cycle of periodic cicadas. Alan Baker writes,

> In each species the nymphal stage remains in the soil for a lengthy period, then the adult cicada emerges after 13 years or 17 years depending on the geographical area. Even more strikingly, this emergence is synchronized among the members of a cicada species in any given area. The adults all emerge within the same few days, they mate, die a few weeks later and then the cycle repeats itself.   [2005, 229]

The question arises: why are the periods prime? Baker [2005] submits that our best explanation appeals to, on the one hand, the evolutionary advantage of minimizing intersection with other cicadas' life cycles, and, on the other, the number-theoretic fact that prime periods minimize this intersection. Moreover, Baker [2005, 233] argues, a number-theoretic result plays a genuinely explanatory—even if not causal—role in this explanation. Letting "lcm" abbreviate least common multiple, the result is: for any prime, p, and for any m,n < p, lcm (p, m) < lcm (n, m).

Baker's argument is questionable. What the above result really seems to explain is *another number-theoretic result*—a result about the primeness of some numbers—not a result about periods of time (Bangu [2008]). But even if we accept it, and accept the Quine–Putnam Thesis* and the *if* direction of the Quine–Harman Thesis*, a remaining problem with Harman's Argument would only be exacerbated. The problem is: *even if the contents of some of our mathematical beliefs are implied by the best explanation of some of our observations, realistically construed, it is hopeless to argue that all are.* Quine and Putnam conceded as much.

3.8 Recreational Mathematics

A well-known piece of set-theoretic folklore says that no empirical scientific theory ever even appears to quantify over objects whose surrogates "live" above rank $\omega + \omega$ in the set-theoretic hierarchy. That is, only a relatively small amount of set theory even appears to be needed to explain our observations. Quine tried to squeeze more out of empirical science. But even he conceded that much modern set theory remains empirically unjustified. He writes,

So much of mathematics as is wanted for use in empirical science is for me on a par with the rest of science. Transfinite ramifications are on the same footing insofar as they come of a simplificatory rounding out, but anything further is on a par with uninterpreted systems.   [1984, 788]

In particular,

I recognize indenumerable infinities only because they are forced on me by the simplest known systematizations of more welcome matters.

Magnitudes in excess of such demands, e.g., Beth$_\omega$ or inaccessible numbers, I look upon only as mathematical recreation and without ontological rights. [1986a, 400]

Recall that Quine is liberal about which beliefs count as empirically justified. He allows that our belief that P may be empirically justified even though P is not "directly" applied by any scientific theory, and even though P plays no causal role in whatever theory incorporates it. Again, the overwhelming majority of our mathematical beliefs had better be inducted (in Russell's sense—see Section 2.4) from what is directly applied if it is to have any hope of being empirically justified, realistically construed. The problem is that even if we induct liberally, we cannot obtain the theories that we actually accept. While the existence of inaccessible numbers is independent of standard set theory (if that is consistent), ZFC, the existence of the cardinal, Beth$_\omega$, is already provable in ZFC.[35] So, Quine must deny that fundamentals of set theory, such as the Axiom of Replacement, are empirically justified. Putnam is even more circumspect. He writes,

> Sets of...cardinality...higher than the continuum...should today be investigated in an "if-then" spirit....[F]or the present we should regard them as what they are – speculative and daring extensions of the basic mathematical apparatus. [1971, 347][36]

Actually, even Putnam's assessment may be too sanguine. Feferman [1992] argues that W, a predicative system based on Weyl [1918], is a mathematically sufficient basis for all of physics. If this is correct, then it shows that still more modern mathematics lacks empirical justification. But Solomon Feferman goes further. He proves that W is proof-theoretically reducible to Peano Arithmetic (PA), and concludes that "the only [mathematical] ontology [that empirical science] commits one to is that which justifies acceptance of PA" [1992, 451]. In particular, "W ... [can be] treated in an instrumental way, its entities outside the natural numbers are regarded as "theoretical"

---

[35] Actually, depending on what takes Beth$_\omega$ to *be*, its existence is provable in ZF without Choice. The $\aleph_\omega$ can be defined in a different way without Choice.

[36] It is undecidable relative to ZFC just how big the cardinality of the continuum is (though it must be uncountable). According to some consistent answers, it is very big. (Indeed, there is no upper bound on how big the cardinality of the continuum can consistently be in ZFC, if ZFC is consistent.) However, such answers are not often endorsed. Putnam appears to change his tune in his [2012] work.

entities, and the justification for its use lies in whatever justification we give to the use of PA" [1992, 451]. Whether or not this is true is certainly debatable. But, if it is true, then, since PA only quantifies over natural numbers, if we accept only its ontology, then we reject the Axiom of Infinity, and with it modern set theory and analysis, realistically construed.

Mark Colyvan dismisses such worries. He suggests that "the higher reaches of set theory, although without physical applications, do carry ontological commitment by virtue of the fact that they have application *in other parts of mathematics*" [2015, § 2, emphasis in original]. But just because some part of mathematics gets applied in empirical science does not mean that all of mathematics is empirically justified! If it did, then our belief in the cardinal, κ, mentioned in Section 2.3, could be empirically justified even if all the mathematics that empirical science ever even appeared to presuppose was Robinson Arithmetic (Peano Arithmetic minus the Induction Schema)! If our set-theoretic belief that P has any hope of being empirically justified, then there must be a *recognizably empirical*—even if not explanatory—argument for P.

Note that our belief that P could be justified in another way. Unlike Quine, a rationalist can hold that our belief in, e.g., Replacement is a priori justified. But such a theorist would be in no position to reject moral realism on the basis of Harman's Argument. Again, if the arguments from the previous chapter are sound, then our moral beliefs have equal claim to being a priori.

The problem is greatly exacerbated if the Quine–Harman Thesis is replaced with the more believable Quine–Harman Thesis*. At most, a small subset of applied mathematics has any claim to being implied by the best explanation of our observations in an explanatory way. Indeed, it is striking that the literature on the enhanced indispensability argument, mentioned in Section 3.7, has been focused on only a few examples. Unsurprisingly, Baker et al. do not purport to establish that our mathematical beliefs generally are empirically justified, realistically construed. They purport to establish that some atomic such beliefs are.[37]

Of course, *if* we grant Harman's Thesis, and assume *both* that the argument from Section 3.6 fails *and* that the *if* direction of the Quine–Harman

[37] More exactly, they are focused on the question of whether there is an empirical—or, anyway, scientific—argument for the existence of mathematical entities. But this is just the question of whether there is such an argument that some atomic (or existentially quantified) mathematical sentences are true, and satisfy Mathematical-Aptness, Mathematical-Truth, and Mathematical-Face-Value (in the sense of Section 1.2).

Thesis* holds, then we could technically conclude that belief in mathematical realism—that is, in the mind-and-language independent truth of *some* atomic mathematical sentences, interpreted at face value—is empirically justified, while belief in moral realism is not.[38] But setting aside the highly conditional nature of this concession, any empirically based "mathematical realism" of this sort would be quite weak. A large—maybe *huge*—part of established mathematics would turn out to be unjustified.

Before concluding, let me highlight a disanalogy with the moral case. While the conjunction of the Quine–Putnam Thesis and the negation of the Quine–Putnam Thesis* has considerable appeal, it is hard to see how anyone could reject Harman's Thesis while accepting the thesis that the contents of our moral beliefs fail to be implied by the best explanation of our observations in an explanatory way. Virtually nobody thinks that *moral* properties just play an indexing role in moral explanations. If alleged moral explanations have any appeal, this is because reference to moral properties seems itself to be somehow explanatory. As Sayre-McCord puts it,

> [C]ertain regularities....are unidentifiable and inexplicable except by appeal to moral properties....Moral explanations allow us to isolate what it is about a person or an action or an institution that leads to its having the effects it does. And these...are uncapturable with finer-grained or differently structured categories.   [1988, 449]

Again, one may doubt that such "moral explanations" are really explanatory. Brian Leiter quips,

> My own feeling is that if I were seeking an explanation for Hitler's conduct [for example] and was offered the explanation "He was morally depraved", I would take such an answer to be a bit of a joke: a repetition of the datum rather than an explanation.   [2001, 91, n. 34]

The point is that there is scant plausibility to the combination of views according to which moral explanations are explanatory, but moral properties fail to play an explanatory role in them.[39]

---

[38] Thanks to Mary Leng for pressing me to make this explicit.
[39] But see Liggins [2014].

3.9 A Priori/A Posteriori Revisited

I have argued that our mathematical beliefs have no better claim to being empirically justified than our moral beliefs, realistically construed. More exactly, I have argued that the range of mathematical beliefs that we actually possess have no such claim. All but one of the arguments for Harman's Thesis work equally as arguments for the view that the contents of our mathematical beliefs fail to be implied by the best explanation of any of our observations. And, even granting Harman's Thesis and the Quine–Putnam Thesis, the Quine–Harman Thesis fails in both directions. At least some of our moral beliefs may be empirically justified because they are based on moral quasi-perceptions, even though none of their contents is implied by the best explanation of any of our observations. And not even Putnam alleges that, if P is implied by the best explanation of our observations, then our belief that P is empirically justified. But if we replace the Quine–Harman Thesis with a more believable thesis, then the argument for parity between morality and mathematics is only strengthened. Finally, even if the Quine–Putnam Thesis, Harman's Thesis, and the Quine–Harman Thesis were all true, it would still not follow that anything like the range of mathematical beliefs that we have is empirically justified.

There is another lesson as well. I have argued that there may be no prin-cipled reason to deny that spontaneous moral judgments count as quasi-perceptions—at least in whatever sense such high-level spontaneous judgments as that such and such people are petting a cat count as quasi-perceptions. Perhaps this shows that such judgments "really are" quasi-perceptions. But perhaps instead it shows that there is no principled distinc-tion between quasi-perception and quasi-intuition. However, a principled a priori/a posteriori distinction would seem to presuppose one. A belief is (defeasibly) a priori justified when it is justified independent of experience—including (quasi)perception, memory, testimony, and introspection.

Actually, similar doubts may even arise for the notions of perception and intuition per se (that is, "successful" quasi-perception and quasi-intuition). One might think that we can perceive ordinary macroscopic objects, like chairs, if we can perceive anything (or, at least, that our spontaneous judg-ments about them may be empirically justified, as discussed in Section 3.6). But the property of being a chair may well be irreducible, and epiphe-nomenal. Suppose that it is. Then the fact that there is a chair in front of us does not cause, and is not even implied by the best explanation of, our perception that this is so. What, in that case, is the important difference between our judgment that there is a chair in front of us and our judgment that what such and such people are doing is wrong? Both judgments may be about concrete things, may involve "spontaneous judgment without any conscious reasoning," and may fail to be fully explained by our conceptual competence. Perhaps there is a principled answer to such questions. But, absent one, there may be yet another basis for skepticism about a useful a priori/a posteriori distinction.[40]

---

[40] Harman himself might say: so much the worse for both judgments! But the idea that our "spontaneous judgment without any conscious reasoning" that there is a chair in front of us is not even *defeasibly* justified if the speculative metaphysical conjecture that the property of being a chair is epiphenomenal happens to be true—unbeknownst to the would-be perceiver!—is surely too much to swallow.

4

Genealogical Debunking Arguments

I have argued that our mathematical beliefs—or anything like the range that
we actually have—possess no better claim to being (defeasibly) justified,
whether a priori or empirically, than our moral beliefs, realistically construed. In particular, Harman's Argument does nothing to show that our
mathematical beliefs possess such a claim. Recall that *Harman's Argument*
includes the following premises:

Quine–Putnam Thesis: The contents of at least some of our mathematical
beliefs, realistically construed, are implied by the best explanation of some
of our observations.

Harman's Thesis: It is not the case that the contents of any of our moral
beliefs, realistically construed, are so implied.

Quine–Harman Thesis: Our belief is empirically justified just in case its
content is implied by the best explanation of some of our observations.

However, Harman's Thesis in tandem with the Quine–Putnam Thesis might
be taken to have a different significance. Even if our moral beliefs have equal
claim to being (defeasibly) justified, perhaps that justification is *defeated* by
knowledge of Harman's Thesis. In particular, perhaps it is defeated by
knowledge that any explanation of our having the moral beliefs that we
have fails to imply their truth, realistically construed. This is a consequence
of Harman's Thesis, given plausible assumptions. By contrast, assuming
the Quine–Putnam Thesis, no analogous argument may be possible in the
mathematical case. It follows that if anyone's moral beliefs are justified,
realistically construed, then this is only because they are ignorant of their
genealogy.

 In this chapter I consider this reasoning in detail. I argue that it misun-
derstands the epistemological significance of indispensability considerations.
Whether the proposition that P is implied by some explanation of our coming to believe that P is indeed predictive of its having epistemically desirable
qualities *when the fact that P would be causally efficacious if it obtained.*

Morality and Mathematics. Justin Clarke-Doane, Oxford University Press (2020). © Justin Clarke-Doane.
DOI: 10.1093/oso/9780198823667.001.0001

But these things are independent when the fact that P would be causally inert, as moral and mathematical facts, realistically construed, are supposed to be. (If they were not supposed to be, then it would not be plausible that any explanation of our having them fails to imply their truth.)

For example, if P would be causally inert if it obtained, then whether the proposition that P is implied by some explanation of our coming to believe that P is independent of whether our belief that P is *sensitive* (that is, roughly, whether had it been that ~P, we would not still have believed that P), *safe* (that is, roughly, whether we could have easily had a false belief as to whether P), or (objectively) probable. I formulate a principle, Modal Security, which constitutes a criterion of adequacy on Genealogical Debunking Arguments (that is, arguments with the aforementioned form). It says that if such arguments undermine, rather than rebut, our target beliefs, then they must at least give us reason to doubt their modal security. But they do not. Whether or not Modal Security is true, however, I argue that Genealogical Debunking Arguments have little force absent an account of the epistemically desirable quality that they are supposed to threaten. I conclude that the real problem to which Genealogical Debunking Arguments point is the reliability challenge, due to Hartry Field. The challenge is to explain the reliability of our moral beliefs, realistically construed. But this challenge has nothing to do with whether the contents of our beliefs are implied by some explanation of our coming to have them. I turn to it in Chapter 5.

4.1 Harman's Thesis as an Underminer

We have seen one way in which Harman's Thesis and the Quine–Putnam Indispensability Thesis might establish a lack of parity between our moral and mathematical beliefs, realistically construed. They might show that our mathematical beliefs are (defeasibly) empirically justified, while our moral beliefs are not, so construed. (More credibly, they might show that there is a dialectically effective argument for the contents of our mathematical beliefs against a thoroughgoing scientific realist that has no analog in the moral case.) But there is another way that Harman's Thesis and the Quine–Putnam Thesis could establish a lack of parity between the cases. Let us suppose, for the sake of argument, that both our moral and mathematical beliefs are (defeasibly) justified somehow or another, realistically construed. Then Harman's Thesis might constitute a *defeater* which, by the Quine–Putnam

Thesis, has no analog in the mathematical case. The reasons to accept Harman's Thesis (surveyed in Section 3.3) seem to be equally reasons to deny that the contents of our moral beliefs are implied by any explanation of our coming to have them, realistically construed. It will be helpful for us to have a name for this particular application of Harman's Thesis. So, let us introduce:

Joyce–Street Thesis: It is not the case that the contents of any of our moral beliefs are implied by any explanation of our coming to have them, realistically construed.[1]

On the other hand, perhaps the primary reason to accept the Quine–Putnam Thesis is that mathematics plays a role in our empirical scientific theories that is similar to the role played in them by their logic. It is a background assumption, so that any such regimented theory will imply a subset of mathematical axioms. Hence, if S is a consequence of that subset, and we believe that S, then S will be implied by any explanation of our coming to believe that S. It will be helpful to have a label for this application of the Quine–Putnam Thesis as well. So, let us introduce:

Steiner's Thesis: The contents of at least some of our mathematical beliefs are implied by every explanation of our coming to have them, realistically construed.

What do I mean by suggesting that Harman's Thesis—and, in particular, the Joyce–Street Thesis—might be a defeater? I mean that our moral beliefs might fail to count as justified given knowledge of it, even assuming that they count as justified absent that knowledge.[2] But evidently the Joyce–Street Thesis does not give us *direct* reason to believe that the contents of our moral beliefs are false, realistically construed. It is not as if the Joyce–Street Thesis is direct reason to believe that slavery is *not* wrong. In the jargon, it is not a *rebutting* defeater. Instead, the Joyce–Street Thesis might give us

---

[1] As in Chapter 3, the rationale for the labels will emerge in what follows. (Note that there is a sense of "explain" in which even debunkers would concede that we can explain our coming to have any belief that we have by appeal to its content (or truth). I can "explain" my coming to believe that P by citing P (or that P is true). That is *why* I believe it. See Fitzpatrick [2016].)

[2] Whether a justified belief, or even an unjustified belief, can render a justified belief unjustified is irrelevant to my purposes here. So, I assume for simplicity that we do not just believe Harman's Thesis, even justifiably. We know it.

reason to give up our moral beliefs, without bearing directly on their contents—somewhat like knowledge that the weatherman is a liar would give us reason to no longer believe that it will rain tomorrow, if we believed this on his say-so, without bearing directly on the weather. It could give us an *undermining* defeater.[3] The idea that the Joyce–Street Thesis constitutes an undermining defeater of our moral beliefs, realistically construed, results in a Genealogical Debunking Argument. And while debunkers focus on evolutionary explanations, this is only for presentational effect (Street [2006, § 11]). (There is another argument to which evolution per se is pertinent. I consider that in Section 5.5.) As Richard Joyce puts it,

> [A]ny epistemological benefit-of-the-doubt that might have been extended to moral beliefs…will be neutralized by the availability of an empirically confirmed moral genealogy that nowhere…presupposes their truth. [2008, 2016]

And Sharon Street remarks,

> Unfortunately for the realist.…[t]o explain why human beings tend to make the [moral] judgments that they we do, we do not need to suppose that these…are *true*. [Street 2008, 208, emphasis in original][4]

So, perhaps knowledge of the Joyce–Street Thesis *undermines* our moral beliefs, realistically construed. Of course, not even debunkers would suggest that it undermines our belief that either slavery is wrong, or it is not the case that it is. Debunkers suggest that it undermines our non-logical moral beliefs—including, crucially, our atomic ones (see, again, Section 1.2).[5]

Street [2006] makes a superficially different suggestion. She says that the Joyce–Street Thesis undermines our belief in moral realism, not our moral beliefs. According to this interpretation, our belief that, say, slavery is wrong is not undermined. Our belief *that the truth of this belief is independent of human minds and languages* is. But, in general, when faced with undermining—or,

---

[3] See Cruz and Pollock [1999, ch. 7, § 2.2] for the distinction between rebutting and undermining (or "undercutting") defeaters.

[4] Street discusses evaluative beliefs in general, but I am presently concerned with the subset of them that are moral beliefs. I return to the topic of evaluative beliefs in general in Chapter 6.

[5] Note that the present sense of "justification" is internalist, not externalist. It concerns our mental states, not our reliability directly. If simple externalism were true, then our beliefs would be justified or not depending on whether they were reliable or not—not depending on whether we had *evidence* that they were reliable or not.

indeed, rebutting—evidence, we do not just reinterpret the contents of our beliefs. We give them up. Consider the weatherman example again. Surely it would be irrational to respond to knowledge that he is a liar by *maintaining* the belief that it will rain tomorrow, but concluding that facts about the rain are mind-and-language dependent. Perhaps matters are different when a whole class of beliefs is on trial—as with the argument, often attributed to Bishop Berkeley,[6] that all of our perceptual beliefs must be about a mental reality, since, otherwise, they would be unjustified. But, if this is so, then there should be an explanation of the difference between the cases. In any event, nothing *non-semantic* turns on whether we say that our moral beliefs are undermined, or instead say that our belief in the independence of their truths is. Roughly, what is at issue between advocates of the above argument is not whether belief in a given *proposition* is undermined. Street herself would agree that our belief in the proposition *that moral realists take "slavery is wrong" to express* is undermined by the above reasoning. She would merely insist that moral realists are wrong about what "slavery is wrong" does express it in natural language. Street and Joyce agree about the non-semantic world.[7]

Genealogical Debunking Arguments are widely supposed not to work equally against mathematical realism. Joyce goes so far as to write:

> [T]he dialectic within which I am working here assumes that if an argument that moral beliefs are unjustified or false would by the same logic show that believing that $1 + 1 = 2$ is unjustified or false, this would count as a *reductio ad absurdum*.   [2007, 182, n. 5]

The reason for the lack of parity stems from the Quine–Putnam Thesis—and, in particular, from Steiner's Thesis. Mark Steiner writes,

> [S]uppose that we believe…the axioms…of number theory…. [S]omething is causally responsible for our belief, and there exists a theory— actual or possible, known or unknown— which can satisfactorily explain our belief

---

[6] See his *Treatise concerning the Principles of Human Knowledge* and *Three Dialogues between Hylas and Philonous*.

[7] Since Joyce should agree that our belief in the proposition that Street takes that sentence to express is *not* undermined. (This is a slight oversimplification. For Joyce and Street alike take "slavery is wrong" to express the proposition *that slavery is wrong*! But one can make all manner of verbal disagreements appear substantive by "semantically descending" in this way. More on this in Section 6.3.)

in causal style. This theory, like all others, *will contain the axioms of number theory....*   [1973, 61, italics in original][8]

Speaking, in particular, of the evolutionary explanation of our coming to have the arithmetic beliefs we have, Joyce appears to apply Steiner's reasoning. He writes,

There is...evidence that the distinct genealogy of [mathematical] beliefs can be pushed right back into evolutionary history. Would the fact that we have such a genealogical explanation of...'1 + 1 = 2' serve to demonstrate that we are unjustified in holding it? Surely not, for we have no grasp of how this belief might have enhanced reproductive fitness independent of assuming its truth.   [2007, 182]

Street makes a similar point with different examples. She writes,

[Genealogical Debunking Arguments do not] go through against realism about non-evaluative facts....In order to explain why it proved advantageous to form judgements about the presence of fires, predators, and cliffs, one will need to posit in one's best explanation that there *were indeed* fires, predators, and cliffs....   [2006, 160–1, n. 35, italics in original][9]

4.2   Three Constraints

Although we have seen ways in which Harman's Thesis and the Quine–Putnam Thesis—and, hence, the Joyce–Street Thesis and Steiner's Thesis—could be challenged, for the present let us grant that they are all true. Then the key epistemological question surrounding Genealogical Debunking Arguments is the following. *How* could knowledge of the Joyce–Street Thesis undermine our moral beliefs, realistically construed?

---

[8] Steiner's reasons for claiming this are different from the ones I gave in Section 3.5 (that a fragment of arithmetic is indispensable to metalogic, which is presupposed by every explanation at all). But this is irrelevant.

[9] I am not suggesting that Street takes numbers to be ontologically like fires or predators. I am saying that the reason she cites for thinking that debunking arguments aimed at realism about the likes of fires or predators do not work is that we must assume the truth of our beliefs about them in the evolutionary explanation of our coming to have those beliefs—a point that we are assuming, with Joyce, holds equally in the mathematical case.

There is a familiar answer, pressed in Street [2006]. Knowledge of the Joyce–Street Thesis gives us reason to believe that it would be a *coincidence* if our moral beliefs were true, realistically construed. Street writes,

> [T]he realist must hold that an astonishing coincidence took place…that as a matter of sheer luck, [causal] pressures affected our…attitudes in such a way that they just happened to land on…the true [moral] views….
>
> [2008: 208–9, emphasis in original]

How does this follow? Street's intuition is that if the Joyce–Street Thesis is true, then the genealogy of our moral beliefs has nothing to do with their truth, realistically construed ([2006, 121]). The forces responsible for our them are independent of their truth, so construed. So, if our moral beliefs are true, realistically construed, then this could only be thanks to a coincidence.

Of course, coincidences happen. But if forced to choose between believing in a coincidence and giving up belief in the realist's truths, Street suggests that we ought to do the latter. As Hartry Field writes in another context, "our belief in a theory should be undermined if the theory requires that it would be a huge coincidence if what we believed about its subject matter were correct" [2005, 77].

There is a problem, however. There are many senses in which it might be said that the truth of our beliefs of a kind, realistically construed, is a coincidence (as we will see). But, if Genealogical Debunking Arguments work against moral realism, but not against mathematical realism, then the present sense of "coincidence" must satisfy the following three conditions:

Moral Debunking: Knowledge of the Joyce–Street Thesis gives us reason to believe that it is a coincidence if any of our moral beliefs are true, realistically construed.

Mathematical Vindication: Knowledge of Steiner's Thesis gives us reason to believe that it is not a coincidence if at least some of our mathematical beliefs are true, realistically construed.

Debunkers' Thesis: If it appears that it would be a coincidence if any of our beliefs of a kind are true, realistically construed, then this *undermines* them, so construed.

What sense of "coincidence," if any, satisfies the above constraints? There are many professed answers to this question. Guy Kahane tells us that, if

Harman's Thesis is true, then "[e]volution is not a truth-tracking process with respect to evaluative truth," and he would presumably say the opposite of mathematical truth under the assumption of Steiner's Thesis [Kahane 2011, 111]. Street says that "most of our evaluative judgements are off track due to the distorting pressure of Darwinian forces," but this is not so of judgments whose truth is implied the evolutionary explanation of our coming to have them [Street 2006, 109]. And Joyce contends, somewhat more precisely, that "moral judgments are the output of a non-truth-tracking process... [where] the intuition at the heart of truth-tracking is that beliefs may or may not be *sensitive* to the facts which they represent" [Joyce 2016, 147, emphasis in original]. But, jargon aside, what is the content of such assertions? What, if anything, could "truth-tracking," "off-track," and "sensitive" *mean* such that the above three theses are all plausible when "coincidence" is analyzed in terms of them? In what follows, I will argue that there is not a satisfying answer.

4.3 Sensitivity

The most obvious way to understand "coincidence" in accord with Joyce's gloss above. Our moral belief that P is true by coincidence if it is not *sensitive*, or does not *track the truth*, in the sense of Nozick [1981, 179]. That is, had it been the case that ~P, we still would have believed that P (had we used the method that we actually used to determine whether P).[10] This reading is suggested by nearly every debunker in at least one passage. For example, Sinnott-Armstrong writes, "[t]he evolutionary explanations [of our having the moral beliefs that we have] work *even if there are no moral facts at all*"[2006, 46, italics mine], and Michael Ruse writes,

> You would believe what you do about right and wrong, irrespective of whether or not a 'true' right and wrong existed! The Darwinian claims that his/her theory gives an entire analysis of our moral sentiments. Nothing more is needed. Given two worlds, identical except that one has an objective morality and the other does not, the humans therein would think and act in exactly the same ways.   [1986, 254]

---

[10] Such conditions concern individual beliefs, and certainly moral realism would not be undermined by evidence that one of our moral beliefs is not sensitive, realistically construed. But if our atomic moral beliefs turned out to be systematically insensitive, realistically construed, and apparent insensitivity is undermining, then this would undermine our belief in moral realism, as defined in Section 1.2. I will not continue to add this qualification.

In an earlier book, Joyce presents a similar thought experiment:

> Suppose that the actual world contains real categorical requirements—the kind that would be necessary to render moral discourse true. In such a world humans will be disposed to make moral judgments...for natural selection will make it so. Now imagine instead that the actual world contains no such requirements at all—nothing to make moral discourse true. In such a world, humans will *still* be disposed to make these judgments...for natural selection will make it so...[D]oes the truth of moral judgments...play a role in their usefulness?....I believe the answer is "No." [2001, 163, italics in original][11]

Strictly speaking, Ruse and Joyce are not concerned with sensitivity here, which operates at the level of individual beliefs. They are concerned with something more like skepticism about the external world. They are asking what we would have believed had the contents of our atomic moral beliefs been systematically false. It could be that, for any one of our atomic moral beliefs, that P, had it been that ~P, we would not have believed that P, even while it is also true that had there been no atomic moral truths at all, we would have believed that P. It could be that the closest worlds in which ~P is a world in which there are atomic moral truths. As Nozick [1981, 227–9] emphasizes, we would have believed that we have hands had there been no atomic perceptual truths—that is, truths about the objects of ordinary perception—too. The best that we can hope for, it seems, is that the closest worlds in which we do not have hands are worlds in which we do not believe that we do (using the method that we actually used to determine this).

Although the question will require revisiting in Section 5.7, it is not preposterous to think that knowledge that our belief is insensitive undermines it—that is, that the present sense of "coincidence" satisfies *Debunkers' Thesis*. For example, suppose that a machine enumerates geographical trivia, deeming its outputs true or false. Independent investigation has confirmed its outputs prior to the last five. The last five outputs are "true." We have deferred to the machine's last five outputs only, and had no prior beliefs of the sort. Today a trusted source tells us that the machine was stuck on "true" in the last five instances. Call this evidence E. Then E does seem to undermine our beliefs in the last five outputs of the machine. But E is not reason to believe that the machine could have easily produced different outputs in

---

[11] We need not assume, with Joyce, that morality presupposes categorical requirements.

the last five cases. That is the point of calling it "stuck." Rather, it is reason to think that, no matter the truth-value of the last five outputs, it still would have called them "true." In other words, E is reason to doubt that our last five geographic beliefs are sensitive. Arguably, this is how E undermines them.

But does the Joyce–Street Thesis *give us reason* to doubt that our moral beliefs are sensitive—that is, is *Moral Debunking* true? Certainly not if the beliefs in question are atomic ones (Sturgeon [1984]). Consider any atomic moral truth, A is M, where A names a particular person, action, or event and M ascribes a moral property. Since the Joyce–Street Thesis does not rebut our moral beliefs, it gives us reason to doubt that *if they are true,* then they are sensitive, if it gives us reason to doubt that they are sensitive.[12] But for any atomic moral belief that A is M, had A not been M, then A would have been different in non-moral respects—since worlds in which the explanatorily basic truths which fix the supervenience of the moral on the non-moral are different and A is not M are presumably more distant from the actual world than worlds in which those truths are the same and A is not M (whatever the exact modal strength of the supervenience).[13] Moreover, had A been different in non-moral respects, then our moral beliefs would have reflected the difference. Had, for example, the people in Harman's example of Section 3.6 been petting the cat rather than pouring gasoline on it and igniting it, we would not have judged that they were doing something wrong to the cat (using the method that we actually used to determine this). This is so whether or not the Joyce–Street Thesis is true.

It is more doubtful that our explanatorily basic moral beliefs themselves are sensitive. A realist might argue that our beliefs in such truths are

---

[12] Suppose that the Joyce–Street Thesis gives us (direct) reason to doubt that *had it been that ~(A is M), we would not have believed that A is M* (using the method that we actually used), but that it does not give us reason to doubt *that A is M.* The proposition that *that it is not the case that had it been that ~(A is M), then we would not have believed that A is M* is true if the embedded counterfactual is false. The embedded counterfactual is false, in turn, if the closest worlds in which ~(A is M) are worlds in which we believe that A is M. The actual world would be such a world if ~(A is M), since we actually believe that A is M. But, by assumption, Harman's Thesis does not give us (direct) reason to believe that ~(A is M). So, if Harman's Thesis gives us reason to think that the closest worlds in which ~(A is M) are worlds in which we believe that A is M, then it gives us reason to believe that, if A is M, this is so.

[13] Note that the explanatorily basic moral truths could be highly disjunctive, if particularists are right (Rosen [Manuscript, 14, n. 3]). That the moral supervenes on the non-moral has been questioned, though even a prominent skeptic labels it "the Least Controversial Thesis in Metaethics" (see Rosen [Manuscript]). If denying supervenience threatened the argument to follow that our beliefs in explanatorily basic moral beliefs (as opposed to our atomic ones) are insensitive, then this would only strengthen my conclusion that the Joyce–Street Thesis cannot undermine our moral beliefs by showing that they are insensitive.

vacuously sensitive on the grounds that any world in which the explanatorily basic moral truths are false is impossible. But, as I argued in Section 3.4, such arguments are fallacious, even assuming a standard semantics for counterfactuals. They assume that metaphysical necessity, as that term is ordinarily used, is the most inclusive objective notion of necessity. Any of various notions of logical possibility afford a counterexample. Moreover, supposing that, for example, Mill's principle of utility is such a truth, and that we believe it, it *does* appear that had it been false, we still would have believed it.

What matters for Genealogical Debunking Arguments, however, is not whether our explanatorily basic moral beliefs are sensitive, but *whether the Joyce–Street Thesis gives us reason to believe that they are not*. And it does not. For instance, if the explanatorily basic moral truths were themselves supervenient with respect to a maximal notion of necessity on truths with which our beliefs co-varied, then, had the explanatorily basic moral truths been different, our explanatorily basic moral beliefs would have been too—even assuming the Joyce–Street Thesis.

Moreover, the converse is also true, so *Mathematical Vindication* is false. Steiner's Thesis does nothing to show that our mathematical beliefs *are* sensitive. On the contrary, we saw that a major motivation for instrumental fictionalism is that "[i]f there were never any such things as [mathematical] objects, the physical world would be exactly as it is right now" [Balaguer 1999, 113]. In particular, "we would have had exactly the same mathematical… beliefs even if the mathematical…truths were different…" [Field 2005, 81]. This is not to concede that our mathematical beliefs are *not* sensitive. I will consider the question in detail in Section 5.6. The point is that merely arguing that the contents of our mathematical beliefs, realistically construed, are implied by some explanation of our coming to have them does nothing to show that they *are* sensitive. It gives "no sense to the idea that if the…facts had been different then our…beliefs would have been different too" [Field 1996, 371]. If it did, then *all* of our true logical beliefs would be *trivially* sensitive, simply because their contents are implied by everything whatever![14]

So, even if debunkers are right that knowledge that our beliefs of a kind, F, are insensitive undermines them (something that I will challenge in

---

[14] In this case, one might retort that our beliefs *are* vacuously sensitive because the logical truths *are* maximally necessary. While I disagree (see my [2018] work), the present point is that showing that the contents of our true logical beliefs are implied by some explanation of our coming to have them is neither here nor there.

Section 5.7), they are mistaken to think that whether our F-beliefs are sensitive depends on whether their contents are implied by some explanation of our coming to have them. Our moral beliefs may be sensitive, even if the Joyce–Street Thesis is true, and our mathematical beliefs may fail to be, even if Steiner's Thesis is true.

4.4 Safety

Another way to understand the relevant sense of "coincidence" is in terms of epistemologists' notion of *safety*. Our belief that P is safe, at first approximation, if it could not have easily been false (using the method that we actually used to determine whether P). But when the truths in question are modally robust, this can only be an approximation. If it could not have easily been that ~P, then, trivially, our belief *that P* could not have easily been false (no matter the method we used to determine whether P). But there is an easy amendment. As Duncan Pritchard writes, "all we need to do is to talk of the doxastic result of the target belief-forming process, whatever that might be, and not focus solely on the belief in the target proposition" [2009, 34]. So, we can say that our belief that P is safe just in case we could not have easily had a false belief as to *whether or not P* (using the method that we actually used to determine whether P). While sensitivity only involves a variation of the truths, safety can involve a variation in our beliefs too.

The idea that our moral beliefs are the products of contingent forces is longstanding. Charles Darwin writes,

> In the same manner as various animals have some sense of beauty, though they admire widely different objects, so they might have a sense of right and wrong, though led by it to follow widely different lines of conduct. If, for instance…men were reared under precisely the same conditions as hive-bees, there can hardly be a doubt that our unmarried females would, like the worker-bees, think it a sacred duty to kill their brothers, and mothers would strive to kill their fertile daughters, and no one would think of interfering. Nevertheless the bee, or any other social animal, would in our supposed case gain…some feeling of right and wrong, or a conscience.   [1871, 70]

Strictly speaking, Darwin does not suggest that we could have easily had different, and so false, moral beliefs. He merely suggests, as Ruse puts it, that "[h]ad evolution taken us down another path, we might well think moral that which we now find horrific, and conversely" [1986, 254]. But this counterfactual is a trivial consequence of Moral-Independence (see Section 1.2). It says that the moral truths do not counterfactually depend on our moral beliefs. No moral realist should deny this. The real worry in the neighborhood is not just that evolution could have led us to have different, as so false, moral beliefs. It is that evolution could have *easily* done this.

Evidence that our belief is unsafe does often seem to undermine it—that is, the present sense of "coincidence" may satisfy *Debunkers' Thesis*. Suppose, for instance, that instead of learning geographical trivia by means of the machine described above, we have had that trivia in our heads for as long as we can remember. We do not recall how we acquired it. But we have no particular reason to doubt it either. Until, that is, we get evidence, E, that when we were very young, we were slipped a pill that induces all but random beliefs in geographical trivia. Our beliefs to do with country capitals, metro populations, and so forth, turn out to be the products of that pill. Then, to the extent that we lack independent evidence to corroborate those beliefs, they would seem to be undermined by our discovery. But E is not evidence that had the truths been otherwise, we still would have believed as we did. Perhaps the closest worlds in which the truths are otherwise are worlds in which the pill gave rise to different beliefs.[15] Instead, E appears to be evidence that we could have easily had opposite such beliefs. Since the geographical truths are not so contingent, this is evidence that we could have easily had false such beliefs—that is, that those beliefs are unsafe. And, arguably, that is how E undermines our beliefs in the example.

However, again, our question is not whether we could have easily had false moral beliefs, but whether the Joyce–Street Thesis *gives us reason to believe that we could have*—that is, whether *Moral Debunking* is true. On the contrary, standard presentations of Genealogical Debunking Arguments themselves illustrate that the Joyce–Street Thesis does not give us reason to believe this. Consider, for example, the most austere interpretation of Street's proposal that "among our most deeply and widely held judgments, we observe many...with exactly the sort of content one would expect if the content of our [moral] judgments had been heavily influenced by selective pressures" [2006, 116]. Suppose that we were evolutionarily bound to have

---

[15] If the pill programmed fixed beliefs, then the undermining could be explained in terms of sensitivity. We would learn that, for any one of the beliefs' contents, P, had it been that ~P, we still would have believed that P.

the explanatorily basic moral beliefs that we have (using the method that we actually used) "for reasons that have nothing to do with their truth." Then we could not have easily had opposite such beliefs. So, given that our moral beliefs are actually true[16], and that the explanatorily basic moral truths could not have easily been different, we could not have easily had *false* such moral beliefs. If our abductive methodology is also "safe"—something debunkers, *qua* scientific realists, would presumably not deny—then we could not have easily had false moral beliefs generally.

I am certainly *not* suggesting that our moral beliefs are safe—or even that we should read Street as suggesting that we could not have easily had different ones. I will ultimately argue for the opposite conclusion. My point is that the *Joyce–Street Thesis gives us no reason to believe that they are not safe.* For all that it says, we could be hard-wired to have the moral beliefs that we do have. And, if we were, we could not have easily had different, and so false, such beliefs.

Conversely, Steiner's Thesis does nothing to show that our mathematical beliefs *are* safe—that is, *Mathematical Vindication* is false. Consider the appeal of so-called *pluralist* views in the philosophy of mathematics (briefly described in Section 1.6, and to be discussed at length in Chapter 6). At first approximation, according to the mathematical pluralist, every consistent mathematical theory is true of its intended subject, independent of human minds and languages. Mark Balaguer writes,

> [Pluralism] eliminates the mystery of how human beings could attain knowledge of mathematical objects. For if [pluralism] is correct, then all we have to do in order to attain such knowledge is conceptualize, or think about, or even "dream up," a mathematical object. Whatever we come up with, so long as it is consistent, we will have formed an accurate representation of some mathematical object, because, according to [pluralism], all [logically] possible mathematical objects exist.   [1995, 317]

This argument does not depend on the existence of a non-mathematical surrogate for every explanation of having the mathematical beliefs that we have. If such surrogates existed for all empirical theories, then it is hard to see what could favor pluralism over fictionalism. Rather, pluralism is

---

[16] This assumption is permissible because the Joyce-Street Thesis is supposed to undermine, rather than rebut, our moral beliefs. So, if the Joyce-Street Thesis gives us reason to doubt that our moral beliefs are safe, then it gives us reason to doubt that if they are true, then they are safe. See, again, Section 4.3.

supposed to resolve a mystery surrounding mathematical knowledge *even under the assumption that mathematics is indispensable to empirical science.* But note that it does nothing to establish that our mathematical beliefs are sensitive.[17] If it is true that, had the objectivist's "universe" been different, our mathematical beliefs would have been the same, then it seems equally true that, had the mathematical pluralist's "pluriverse" been different, our mathematical beliefs would have been the same. If pluralism guards against any kind of mismatch between our mathematical beliefs and the truths, then it guards against our *beliefs being different* while the mathematical truths fail to be.[18] But there is no reason to guard against this mismatch as it stands. Again, it is a trivial consequence of Mathematical-Independence (from Section 1.2). It just says that the mathematical truths do not counterfactually depend on our mathematical beliefs. The problem arises when we add that we could have *easily* had different mathematical beliefs. Given that the mathematical truths could not have easily been different, it follows that we could have easily had *false* such beliefs. By itself, Steiner's Thesis does nothing to show that we could not have. Indeed, if a theorem of arithmetic, S, is a background assumption to the theory that explains your coming to believe that S, then it is equally a background assumption to the theory of my coming to believe that ~S—since it is a background assumption to every explanation.

Again, while debunkers may be right that knowledge that our beliefs of a kind, F, are unsafe undermines those beliefs, they err in assuming that whether our F-beliefs are safe depends on whether their truth is implied by some explanation of our coming to have them. On the contrary, our moral beliefs may be safe even though the Joyce–Street Thesis is true, and our mathematical beliefs may fail to be safe even though Steiner's Thesis is. To sum up: whether the contents of our beliefs of a kind, F, are implied by some explanation of our coming to have them is *independent of the question of whether they are sensitive and safe*, realistically construed.

4.5 "Connection"

If the Joyce–Street Thesis gives us no reason to doubt that our moral beliefs are sensitive and safe, is there any other sense of "coincidence" in which it

---

[17] Contra the apparent suggestion in Field [2005, 78 and 81].
[18] This is an oversimplification, since the pluralist cannot hold that if the *proposition* that P is actually true, then had we believed ~P, we still would have believed truly. See Section 6.1.

shows that the truth of our moral beliefs would be a coincidence, realistically construed? It does not give us reason to doubt that the probability that our moral beliefs are true is high, contra Street [2006, 129].[19] Either the probability in question is epistemic or it is objective. Whether the epistemic probability that our moral beliefs are true is high, given the Joyce–Street Thesis, is just what is in dispute. But the Joyce–Street Thesis does not give us reason to doubt that the *objective* probability that our moral beliefs are true is high. For any explanatorily basic moral truth, P, presumably $Pr(P) = 1$, given that such truths are metaphysically necessary. (We *could* require that $Pr(P) = 1$ only if P is necessary in an even stronger sense—e.g., maximally necessary. But then it would turn out that virtually every proposition of interest, including all atomic mathematical propositions, have equal claim to being objectively improbable. See, again, Section. 3.4.) Also, for all that has been argued, it may be that $Pr(\text{we believe that } P) \simeq 1$. But, then, it may be that $Pr(P \ \& \ \text{we believe that } P) \simeq 1$, by the probability calculus. Since (P & we believe that P) implies (our belief that P is true), it may be that Pr (our belief as to whether P is true) $\simeq 1$ as well.

Nor does Steiner's Thesis give us reason to believe that the objective probability that our mathematical beliefs are true *is* high. Again, take any logical truth that you believe, P. Then P is implied by every explanation, and so is certainly implied by any explanation of your coming to believe that P. The problem is that, while $Pr(P) = 1$, we may have no reason to suspect that $Pr(\text{we believe that } P) \simeq 1$. It could be that you decided to believe P by consulting tea leaves, but were bound to take a stand on the question of whether P. And, yet, if $Pr(\text{we believe that } P)$ is low, then $Pr(P \ \& \ \text{we believe that } P)$ is low, so $Pr(\text{our belief as to whether P is true})$ may be low too.

(There are compelling reasons to believe that an analysis of "coincidence" in terms of objective probability would not satisfy *Debunkers' Thesis* anyway. We seem to be able to know that P even when we also know that whether our belief that P is true is objectively improbable (Williamson [2013]). My point is that, even under the dubious assumption that such an analysis satisfied *Debunkers' Thesis*, it would still not satisfy *Moral Debunking* and *Mathematical Vindication*.)

Perhaps we should understand "coincidence" causally? If we did, then *Mathematical Vindication* would be false. As we saw in Section 3.3, the whole point of the fictionalist program to purge empirical science of

[19] See my [2016, § 2.6] and Baras [2017] for an argument to this effect.

mathematical sentences, realistically construed, is that there appears to be no causal connection between mathematical facts and anything else, realistically construed. Moreover, few would claim that knowledge that there is no causal connection between our beliefs and the facts undermines them—i.e., that *Debunkers' Thesis* is true under the present interpretation of "coincidence."[20] Even Goldman [1967], the original presentation of the causal theory of knowledge, explicitly rejects a causal theory of a priori knowledge. And Alvin Goldman himself long ago jettisoned the theory in the empirical case as well.

Is there any other sense of "coincidence" such that *Moral Debunking*, *Mathematical Vindication*, and *Debunkers' Thesis* are all true? It is sometimes suggested that we should appeal to hyperintensional ideology, like grounding or constitution (Chudnoff [2013, Pt III], Bengson [2015]). Suppose that we believe that P on the basis of an intuition that P. Then perhaps our (token) intuition that P can only be partly constituted by or grounded in the fact that P if P is implied by some explanation of our coming to believe that P. And maybe our belief that P is true by coincidence if it is based on intuition but not so constituted or grounded. After all, in order to know that P *via* perception, it is not enough to have a veridical quasi-perception that P. The quasi-perception must have been (appropriately) caused by the fact that P.[21] So maybe, in a similar way, in order to know that P by intuition, it is not enough to have a veridical "quasi-intuition" that P. The quasi-intuition must be constituted by or grounded in the fact that P.

I will not argue about the requirements on knowledge. What matters here is justification. And while we are free to use "coincidence" so as to satisfy the above condition, there seems to be no reason to think that *Debunkers' Thesis* is true of it. Assuming that our allegedly deficient quasi-intuition that P has the content that P (and so is veridical), it is hard to see why the extra constitution or grounding connection should be of any epistemological concern. *While causal connection is predictive of co-variation with the truth, a constitutive or grounding connection is not.* The assumption that our intuition that P was constituted by or grounded in the fact that P no more shows that our belief that P is sensitive, safe, or (objectively) probable than does the assumption that P was implied by some explanation of

---

[20] The only possible exception I know of is Colin Cheyne. He holds that, "[i]f Fs are noncomparative objects, then we cannot know that Fs exist unless our belief in their existence is caused by: (a) an event in which Fs participate, or (b) events in which each of the robust constituents of Fs participate, or (c) an event which proximately causes an F to exist" [Cheyne 1998, 46].
[21] Recall that a quasi-perception is just like a perception, but not factive. See, again, Section 3.6.

our coming to believe that P. From anything resembling the standpoint of trying to have true beliefs, it would surely be preferable to have a safe, sensitive, and objectively probable quasi-intuition that P that is not "connected" to the fact that P than to have an unsafe, insensitive, and objectively improbable intuition that is.[22]

4.6   Modal Security

The above considerations suggest the following constraint on undermining evidence (introduced in slightly different form in Clarke-Doane [2015a]):

Modal Security: If evidence, E, *undermines* (rather than rebuts) our belief that P, then E gives us (direct) reason to doubt that our belief is sensitive or safe (using the method that we actually used to determine whether P).[23]

The intuition behind Modal Security is that if our belief that P is (defeasibly) justified, and we get evidence that neither tells directly against its content nor against the security of its truth, then it is irrational to give it up. Modal Security does not give an analysis of "undermines," since it does not even give a sufficient condition for being an underminer. It allows that knowledge that our belief is neither safe nor sensitive is *not* undermining. It merely puts a constraint on underminers. *If* E undermines our belief, *then* E must be (direct) reason to doubt its Modal Security. Modal Security says that, insofar as the notion of undermining (rather than rebutting) evidence makes sense, evidence cannot undermine our belief without telling against its modal security.

It might be thought that there is a large class of counterexamples to Modal Security. Take any metaphysical necessity, P, that we justifiably believe. Now consider evidence, E, that we were bound to believe it. For instance, let us replace "geographical" with "mathematical" in the trivia

---

[22] Nor would it help to claim that knowledge is the norm of belief. Even if we treat it that way, maybe we should not. So treating it would seem to be a bad policy, assuming that knowledge can come apart from justified true belief that is sensitive, safe, and (objectively) probable. For an argument that we should not treat knowledge as the norm of belief—indeed, should not care much about knowledge at all—see Papineau [2019].

[23] Note that Modal Security does not include a probabilistic condition. My argument would work equally if one were added. Again, the Joyce–Street Thesis gives us no reason to doubt that the probability that our moral beliefs are true is high. But, as I explained in Section 4.6, I do not believe that a probabilistic analysis of "coincidence" satisfies *Debunkers' Thesis*.

machine example from Section 4.3, where E is evidence that the machine was stuck in the last five instances. Then surely E undermines our belief in its last five outputs, whether they concern geography or numbers. But if E is reason to doubt that our belief that P is sensitive, it is equally reason to think that it *is* sensitive. Had it been that ~P, anything goes—since P is necessary.

But this argument just replays the mistake with "necessary" that I identified in Section 3.4. The sense in which the mathematical truths are necessary has no claim to being maximal. So, E may undermine our belief that P by giving us reason to doubt that it is sensitive—just as in the geography case.

However, while Modal Security is consistent with this diagnosis, it turns out to be unwise, for reasons that will emerge in Section 5.7.[24] What the case actually shows is that our definition of safety is still not quite adequate. One's belief should only count as safe if "one avoids false belief in *every case similar enough* to [P]" [Williamson 2000, 124, my emphasis]. In other words, our belief that P is *not* safe if we could have easily had a false belief as to whether Q, where Q is any proposition similar enough to P (using the method that we actually used to determine whether P)—even if we could not have easily had a false belief as to whether P per se (using that method). What counts as similar enough? The answer is vague and context-sensitive. But there are clear cases. If the calculating rules I appeal to when computing tips are unreliable in general, but give the right answer to the question, e.g., "what is 20 percent of $9.98?," then my belief that 20 percent of $9.98 is $2.00, which I formed using those rules, is unsafe. It does not matter that I could not have easily had a false belief as to *that proposition* using those rules.

The mathematical trivia machine case can now be handled differently. E is evidence that, even if the machine had considered falsities last, it still would have called them truths. But it cannot just be this fact which is undermining. If we know that worlds in which it considers falsities last are distant, then evidence that, had we been in one, we would have had false mathematical beliefs would not undermine them. So, it must be added that we know that such worlds are appropriately similar to ours. But if this is added, then E is evidence that our last five mathematical beliefs are unsafe. It is evidence that, for any one of their contents, P, we could have easily had a false belief as to whether Q, where Q is any false proposition similar

---

[24] It also would not handle inevitable belief in *maximally* necessary truths.

enough to P that the machine could have easily considered in the last 10 cases. (The original formulation suffices to cover this case if the machine could have easily considered ¬P in the last ten cases.)

Another kind of potential counterexample to Modal Security concerns misleading higher-order evidence. Suppose, for example, that our belief that P is (defeasibly) justified, but we get evidence that a false theory of justification is true according to which P is not. Perhaps we are Philosophy 101 students and read Descartes on why justification requires infallibility, for instance.[25] Then it might be thought that our belief that P undermined, since it fails to count as justified by the lights of the theory of justification for which we have good evidence. Of course, generally, evidence that our belief that P is not justified *is* evidence that our belief that P is not safe or sensitive (or objectively probable). So what we need to imagine is that we somehow get (direct) evidence that our belief is not justified, *without getting (direct) evidence that it is not safe, sensitive (or objectively probable)*. But then it is difficult to see why we *should* give up that belief. Learning that our belief fails to count as justified in this way seems to be like learning that it is not polite. The truth-seekers among us should out of such epistemic etiquette.[26]

A more serious problem is that we could get *indirect* evidence that our beliefs are unsafe or insensitive (Schechter [2018]). However, Modal Security as formulated implies that such evidence could never undermine. This seems wrong. As Dan Baras and I put it:

Suppose that…your belief that Abe is reliable was fully based on the testimony of Ben….Now suppose you receive a new testimony from Cas, who's otherwise justifiably considered trustworthy, that Ben is unreliable. Now that is not direct reason to doubt that Abe's testimony is safe or sensitive…. [Nor can] it…be a direct reason to doubt that P is true. But it

[25] Thanks to David James Barnett for suggesting this example. (Whether Descartes should actually be read in the way that he tends to be read in Philosophy 101 classes is irrelevant to my purposes here.)
[26] Of course, there is a sense in which even learning that our beliefs are not sensitive or safe (or objectively probable) can appear epistemically irrelevant. If we are justified in believing that P, and gain knowledge that our belief that P is unsafe, say, then there is a temptation to retort: what does that have to do with P? But this is just the puzzle of indirect evidence—evidence which gives us reason to give up our belief without giving us direct reason to believe the negation of its content. *If the notion of indirect evidence makes sense at all* (and it would require a radical revision to our epistemic practices if it did not), then learning that one's method belief is unreliable in the sense of being unsafe or insensitive (or, perhaps, objectively improbable) seems to have claim to undermining.

is surely a defeater of Bp (the belief the p) just as much as the case in which Ben [testifies] that Abe is unreliable.

(Baras and Clarke-Doane [Forthcoming, 16])

There is an easy fix, however. The upshot of such examples is not that underminers can undermine, but not by threatening the modal security of our beliefs. It is that E can undermine our belief that P via a chain of evidence which *bottoms out* in a direct reason to doubt safety or sensitivity (perhaps not of our belief that P). In the case above, Cas's testimony *is* a direct reason to believe that Ben is unreliable. That is a direct reason to believe that any belief based on Ben's testimony is unsafe or insensitive. And the belief formed on the basis of Ben's testimony was that Abe is reliable. So, Cas's testimony undermines the belief that <the belief based on Abe's testimony is safe and sensitive>. That is how it undermines beliefs based on Abe's testimony. We can take this into account by formulating Modal Security as a recursive principle:

Modal Security (Baras and Clarke-Doane [Forthcoming, Section 8]): If evidence, E, *undermines* (rather than rebuts) our belief that P, then E gives us direct reason to doubt that our belief is sensitive or safe (using the method that we actually used to determine whether P) *or* E undermines our belief that <our belief that P is safe and sensitive>.

There are alternatives to Modal Security. John Pollock says that underminers attack the connection between our grounds for belief and the belief itself. If we take this connection to be the *justifying* connection, however, then "underminers must always be evidence that you were never really justified in the first place" [Gibbons 2014, 107]. The whole point of Genealogical Debunking Arguments is that our moral beliefs were justified prior to gaining knowledge of their genealogy. Pollock himself cashes out the connection as follows: "If believing P is a defeasible reason for S to believe Q, M* is an undercutting [undermining] defeater for this reason if and only if M* is a defeater and M* is a reason for S to doubt or deny that P would not be true unless Q were true" [Cruz and Pollock 1999, p. 196]. This condition implies that an undermining defeater must give us reason to doubt the sensitivity of our belief that Q if our belief forming method and the reason to believe Q co-vary. But even if some underminers of this form failed to satisfy Modal Security, this would require an amendment of detail, not spirit. It would not violate the intuition that if E undermines our belief, then it gives us reason to doubt its modal security.

Modal Security is controversial (Berry [Forthcoming], Faraci [Forthcoming], Jonas [2017], Klenk [Manuscript], Locke and Korman [Forthcoming, § 6], Schafer [2017], Schechter [Forthcoming], Tersman [2016], Warren [2017], Woods [2018]). But even if it is false, it is surely tendentious to assume that the Joyce–Street Thesis undermines (rather than rebuts) our moral beliefs, realistically construed, *given that it fails to do so by threatening their modal security (or even objective probability)*. Why, then, have so many philosophers supposed that there is a sense of "coincidence" satisfying *Moral Debunking*, *Mathematical Vindication*, and *Debunkers' Thesis*? The answer, I suspect, is that they have inducted from a special case. When the content of our belief in question, that P, corresponds to a fact which would be causally efficacious if it obtained, then evidence that P is not implied by any explanation of our coming to believe that P is generally evidence that our belief that P is not safe or not sensitive. The problem is that evidence that this is so is *not even defeasible evidence for lack of safety or sensitivity when P corresponds to a causally inert fact*—as debunkers take moral and mathematical facts to be.

Consider our belief that there is a piece of paper in front of us. Although the assumption is questionable (Merricks [2001]), let us assume that the corresponding fact would be causally efficacious, if it obtained. Then, in paradigm cases, the proposition that there is a piece of paper in front of us is implied by some explanation of our coming to believe that there is *just in case* that belief is sensitive and safe (and objectively probable). The truth that there is a piece of paper in front of us plays a *causal role* in a true explanation of our coming to believe that there is. So, for example, had it not been part of that explanation, it seems that we would not have believed that there was a piece of paper in front of us (barring overdetermination). It matters whether the truth that there is a piece of paper in front of us is implied by some explanation our coming to believe this because, *in such cases*, this is *predictive* of epistemically valuable properties—such as sensitivity and safety. However, if the fact that there is a piece of paper in front of us is causally inert because it is *epiphenomenal*, as some metaphysicians allege it would be, then our belief that there is a piece of a paper in front of us will be equally safe and sensitive (and objectively probable), if true, even though the truth of that proposition is not implied by any explanation of our coming to have that belief. Conversely, our belief in a logical truth may be unsafe and insensitive (and objectively improbable) even though it is implied by every explanation at all.[27]

Some might still feel that *connection* per se matters, even if it is not predictive of reliability in any useful sense (Locke and Korman [Forthcoming]). And I suppose that P's being implied by some explanation of our coming to believe that P, even for causally inert P, is *some* kind of connection between P and our (token) belief that P. But, again, while we are free to use "coincidence" to satisfy the condition that P is true by coincidence if all explanations of our coming to believe that P fail to imply that P, there seems to be no epistemic reason to *care* about coincidences, so conceived. To repeat: from anything resembling the standpoint of trying to have true beliefs, it is surely preferable to have a safe, sensitive, and probable belief which is not "connected" to the truth than to have an unsafe, insensitive, and improbable belief that is.[28]

4.7 Conclusions

Genealogical Debunking Arguments are fallacious. They are only defensible if whether our belief that P is implied by some explanation of our coming to believe that P is predictive of whether our belief that P has epistemically valuable properties. But it is not when the facts in question would be causally inert, as debunkers themselves suppose that moral truths would be. (Again, if they did not assume this, then they could not assume the Joyce–Street Thesis, that any explanation of our coming to have the moral beliefs that we have fails to imply their truth.)

However, it does not follow that there is no epistemic mystery surrounding moral knowledge, or even that that mystery has nothing to do with the genealogy of our moral beliefs, in *some* sense of "genealogy." Street alludes to it when she writes, "the realist must hold that an astonishing coincidence took place…that as a matter of sheer luck, [causal] pressures affected our…attitudes in such a way that they just happened to land on…the true [moral] views…." [2008, 208–9]. Her mistake was to think that *whether* it

---

[27] Or, again, if one has qualms about sensitivity as applied to logical truths, the same point holds of the fragment of arithmetic truths that are indispensable to metalogic (see, again, Section 3.5).

[28] Contra the apparent suggestion of Korman and Locke [Forthcoming, 19, n. 30].

would be a massive coincidence that our beliefs of a kind are true has anything to do with whether their contents are implied by some explanation of our coming to have them. As I have argued, these questions are independent. So, in Chapter 5 I set aside the question of whether Harman's Thesis *shows* that it would be a coincidence that our moral beliefs are true, realistically construed, and the question of whether the Quine–Putnam Thesis shows the opposite in the mathematical case. I turn directly to the question of whether it would, in fact, be a coincidence if our moral or mathematical beliefs were true, so construed.

5

Explaining our Reliability

I have argued that our mathematical beliefs have no better claim to being (defeasibly) justified, whether a priori or empirically, than our moral beliefs, realistically construed. Moreover, I have argued that Genealogical Debunking Arguments, as standardly formulated, are fallacious. They assume that whether the truth of our beliefs of a kind, F, is coincidental depends on whether the contents of our F-beliefs are implied by some explanation of our coming to have them. There is, in general, no *epistemologically important* sense of "coincidence" in which this is true.

However, the challenge to explain the reliability of moral and mathematical beliefs, realistically construed (or to show that their truth is no coincidence), remains. In this chapter I substantially clarify this challenge, as outlined in the mathematical case by Hartry Field. I then seek an interpretation of "explain the reliability" which satisfies the following three conditions.

Moral Unreliability: It appears impossible to explain the reliability of our moral beliefs, realistically construed.

Mathematical Reliability: It does not appear impossible to explain the reliability of our mathematical beliefs, realistically construed.

Undermining Inexplicability: If it appears impossible to explain the reliability of our beliefs of a kind, F, realistically construed, then this *undermines* them, so construed.[1]

I consider a wide range of proposals. I begin with Benacerraf's own, and then turn to improvements on it. I argue that, even when the proposals satisfy *Moral Unreliability* and *Mathematical Reliability*, they do not satisfy *Undermining Inexplicability*. I then turn to more promising analyses, in terms of variations of the truths and variations of our beliefs. The best version of the former is the challenge to show that our beliefs are *sensitive*—that is,

---

[1] Note the similarity between the *Undermining Inexplicability* and *Debunkers' Thesis*, from Section 4.2. See Section 5.3 for discussion of the connection between the two.

Morality and Mathematics. Justin Clarke-Doane, Oxford University Press (2020). © Justin Clarke-Doane.
DOI: 10.1093/oso/9780198823667.001.0001

that, for any one of their contents, P, had it been that ~P, we would not still have believed that P (using the method that we actually used to determine whether P). This challenge is widely supposed to admit of an evolutionary answer in the mathematical case, but not in the moral. I argue, on the contrary, that it may admit of an evolutionary answer in the moral case, and not the mathematical. But this is only because the sensitivity challenge is trivial to meet when the truths in question ascribe supervenient properties of concrete things, and impossible to meet when they do not. So, *Undermining Inexplicability* fails for this sense of "explain the reliability." This leaves analyses in terms of the variation of our beliefs. I argue that the best version of these is the challenge to show that our beliefs are safe. Recall that our belief that P is *safe* when we could not have easily had a false belief as to whether Q, for any proposition similar enough to P, Q (using the method that we actually used to determine whether P). Understanding the reliability challenge as the challenge to show that our beliefs are safe explains the widespread conviction that, whatever its costs, the view known as "mathematical pluralism" at least affords an answer to the reliability challenge. This understanding also illuminates the epistemic significance of genealogy and disagreement. I conclude that whether the reliability challenge is equally pressing in the moral and mathematical cases depends on whether realist pluralism is equally viable in metaethics and the philosophy of mathematics.

5.1 Justification and Reliability

In "Mathematical Truth," Paul Benacerraf presented an epistemological problem for mathematical realism. "[S]omething must be said to bridge the chasm, created by ... [a] realistic ... interpretation of mathematical propositions ... and the human knower," he writes. For prima facie, "the connection between the truth conditions for the statements of [our mathematical theories] and ... the people who are supposed to have mathematical knowledge cannot be made out" [1973, 673]. The problem presented by Benacerraf— variously called "the Benacerraf Problem," the "Access Problem," the "Benacerraf–Field Challenge," or, as I will call it (following Schechter [2010]), the "reliability challenge"—has largely shaped the philosophy of mathematics. Realist and anti-realist views have been defined in reaction to it. But the influence of the Benacerraf Problem is not limited to the philosophy of mathematics. The problem is now thought to arise in a host of areas, including metaethics. The following quotations are representative:

The challenge for the moral realist…is to explain how it would be anything more than chance if my moral beliefs were true, given that I do not interact with moral properties.…[T]his problem is not specific to moral knowledge.…Paul Benacerraf originally raised it as a problem about mathematics.   [Huemer (2005, 99][2]

It is a familiar objection to…modal realism that if it were true, then it would not be possible to know any of the facts about what is…possible.…This epistemological objection…may…parallel…Benacerraf's dilemma about mathematical…knowledge. Stalnaker [1996, 39–40]

We are reliable about logic.…This is a striking fact about us, one that stands in need of explanation. But it is not at all clear how to explain it.…This puzzle is akin to the well-known Benacerraf–Field problem.…

[Schechter 2013, 1]

Benacerraf's argument, if cogent, establishes that knowledge of necessary truths is not possible.   [Casullo 2002, 97]

The lack of…an explanation [of our reliability] in the case of intuitions makes a number of people worry about relying on [philosophical] intuitions. (This really is just Benacerraf's worry about mathematical knowledge.)

[Bealer 1999, 52, n. 22]

[W]hat Benacerraf…asserts about mathematical truth applies to any subject matter. The concept of truth, as it is explicated for any given subject matter, must fit into an overall account of knowledge in a way that makes it intelligible how we have the knowledge in that domain that we do have.

[Peacocke 1999, 1–2]

One upshot of the discussion to follow is that even the above understates the case. Even Genealogical Debunking Arguments, discussed in Chapter 4, are best understood as applications of the reliability challenge.

The request for an "account of the link between our cognitive faculties and the objects known" can be interpreted in at least two ways—ways that Benacerraf does not distinguish.[3] First, it can be interpreted as the challenge to

---

[2] John Mackie expressed a similar reservation. He writes, "It would…make a [radical] difference to our epistemology if it had to explain how…objective values are or can be known, and to our philosophical psychology to allow such knowledge" [1977, 24]. "If we were aware [of objective values], it would have to be by some special faculty of moral perception or intuition, utterly different from our ways of knowing everything else" [1977, 38].

[3] Schechter [2006] marks a cognate distinction.

explain the (defeasible) *justification* of our beliefs. Second, it can be interpreted as the challenge to explain their *reliability*. (It can also be understood as the challenge to *justify*, in a dialectical sense, our beliefs, as described in Section 3.8 (Field [1989, 26]).) We have already seen that the moral realist seems equally placed to answer the first challenge. But an answer to the first challenge does not translate into an answer to the second.

Consider, again, Kurt Godel's suggestion that, "despite their remoteness from sense experience, we...have something like a perception also of the objects of set theory as is seen from the fact that the axioms force themselves upon us as being true" [1947/1983, 483–4]. Godel complains, "I don't see any reason why we should have less confidence in this kind of perception, that is, mathematical intuition, than in sense perception..." [1947/1983, 483–4]. Godel's remarks are widely regarded as being unresponsive to part of Benacerraf's epistemological challenge. Let us grant that appeals to intuition help to explain the (defeasible) justification of our mathematical beliefs.[4] They help to explain why it would be reasonable to trust their contents, absent reason to doubt them. Perhaps intuitions have a similar phenomenology as perceptions. So, absent reason to doubt their contents, they justify them if perceptions justify theirs (Bengson [2015], Chudnoff [2013, Pt II]). Still, there is surely something mysterious about Godel's epistemology. Why is being the content of an intuition a reliable symptom of being true? As Benacerraf puts it,

> What troubles me is that without an account of *how* the axioms "force themselves upon us as being true," the analogy with sense perception and physical science is without much content....In physical science we have at least a start of...an account [of the link between our cognitive faculties and the objects known].... To be sure, there is a *superficial* analogy.... [W]e "verify" axioms by deducing consequences from them concerning areas in which we seem to have more direct "perception" (clearer intuitions). But we are never told how we know even these, clearer, propositions.
>
> [1973, 674, italics in original]

Despite common suggestions to the contrary, a similar point applies to W. V. O. Quine's empiricist epistemology of mathematics (discussed in Section 3.2). Russell Marcus writes,

---

[4] Thus bracketing the fact that, as we saw in Chapter 2, judgments regarding what "forces" itself on us as being true are far from uniform, even among experts. (Thanks to Katja Vogt for reminding me to flag this.)

It is one of Quine's great achievements to notice that the access problem in the philosophy of mathematics becomes obsolete once we recognize that ontological commitment is a matter of formulating theories rather than grounding each individual claim in sense experience or rational insight.

[2015, 51]

Quine, recall, holds that electrons and numbers are on an epistemic par. But this is ambiguous. It could mean that what explains the justification of our belief in electrons is the same as what explains our justification for belief in numbers—or, following Hilary Putnam, that we can (dialectically) justify these beliefs in the same way. In both cases, perhaps, the content of the belief is implied by the best explanation of our observations. Alternatively, it could mean that what explains the reliability of our beliefs about electrons is the same as what explains the reliability of our beliefs about numbers. Only the latter is pertinent. But given the nature of mathematical entities, there is no apparent reason to suppose that, just because truths about numbers are implied by the best explanation of our observations, those observations are responsive to the truths about numbers in the way that they are responsive to the truths about electrons (Section 3.4). Again, unlike electrons, numbers are apparently causally inert.[5]

5.2 Clarifying the Challenge

Let us call the challenge to explain the reliability of our beliefs of a kind, F, realistically construed, the *reliability challenge* for F-realism. Then, although Benacerraf first drew attention to a challenge in the vicinity, the canonical presentation of the reliability challenge is actually due to Hartry Field (Liggins [2010], Linnebo [2006]). The guiding idea is that:

[O]ur belief in a theory should be undermined if the theory requires that it would be a huge coincidence if what we believed about its subject matter were correct. But mathematical theories, taken at face value, postulate mathematical objects that are mind-independent and bear no causal or

---

[5] Colyvan appears to make the same mistake, despite suggestions to the contrary. He writes, "[L]et's take a...charitable reading of the...[Benacerraf] challenge, according to which the challenge is to explain the reliability of our systems of beliefs ....Once the challenge is put this way, we see that Quine has already answered it: we justify our *system of beliefs* by testing it against *bodies of empirical evidence*" [2007, 111, emphasis in original]. See also Hart [1996].

spatiotemporal relations to us, or any other kinds of relations to us that would explain why our beliefs about them tend to be correct; it seems hard to give any account of our beliefs about these mathematical objects that doesn't make the correctness of the beliefs a huge coincidence.   [Field 2005, 77]

This is reminiscent of Sharon Street's suggestion that:

[T]he realist must hold that an astonishing coincidence took place – claiming that as a matter of sheer luck... [causal] pressures affected our evaluative attitudes in such a way that they just happened to land on or near the true [moral] views among all the conceptually possible ones.

[2008, 208][6]

Field gives his most precise statement of the challenge in his [1989, Introduction]. He writes,

We start out by assuming the existence of mathematical entities that obey the standard mathematical theories; we grant also that there may be positive reasons for believing in those entities.... But Benacerraf's challenge... is to... explain how our beliefs about these remote entities can so well reflect the facts about them.... *[I]f it appears in principle impossible to explain this*, then that tends to *undermine* the belief in mathematical entities, *despite* whatever reason we might have for believing in them.

[Field 1989, 26, emphasis in original]

So formulated, the reliability challenge for realism about an area, F, has a number of virtues (some of which Field does not seem to recognize). First, it cannot be dismissed as a puzzle of no practical significance. The apparent impossibility of answering the challenge is supposed to *undermine* our F-beliefs, realistically construed. If it appears impossible to answer, then we ought to *change our F-beliefs*, so construed.[7] The challenge might not have this significance if it merely purported to show, as Benacerraf suggests in the mathematical case, that our F-beliefs fail to qualify as *knowledge*, realistically

---

[6] Enoch [2011] emphasizes the similarities between Field's and Street's challenges, at this level of abstraction. (Again, Street is focused on evaluative beliefs generally. I return to them in Chapter 6.)

[7] Some philosophers claim that it only follows that we ought to change our belief in math-ematical realism, but not our mathematical beliefs. Again, if this suggestion is coherent, the point is merely verbal. See Section 4.1.

construed. If our F-beliefs are justified, so construed, and we can explain their reliability, then who cares if they fail to qualify as knowledge?[8]

The second virtue of Field's formulation is that it is not a "convince the skeptic" challenge.[9] Field *grants* that our F-beliefs are (actually) true and (defeasibly) justified, realistically construed. His contention is that it appears impossible to explain their reliability, *even granted these assumptions*. If Field did not grant these things, then his challenge would overgeneralize.[10] The evolutionary explanations of our having reliable mechanisms for perceptual belief, and the neurophysical explanations of how those mechanisms work such that they are reliable, all presuppose the (actual) truth of our perceptual beliefs.[11] These explanations do not *state* that our perceptual beliefs are true. But if we were not justified in believing that our perceptual beliefs were true, then we would not be justified in believing the evidence to which the explanations of the reliability of our perceptual faculties appeal. Field's contention is that there is an epistemic *difference* between our mathematical beliefs and our perceptual beliefs. Street makes a similar caveat when she writes that "the challenge doesn't go through against realism about non-evaluative facts such as facts about fires, predators, cliffs, and so on" [2006, 160, n. 35].

Third, Field's challenge is distinct from the challenge to explain the *determinacy* of our F-beliefs (Putnam [1980], Field [1989, Introduction]). In fact, although this has not been widely recognized, the challenges are in tension. According to the challenge to explain the determinacy of our mathematical beliefs, there is nothing in our practices, or in the world, that could pin down an intended model of set theory. We know, for instance, that if ZFC is consistent, then so is ZFC + Cantor's Continuum Hypothesis (CH) and ZFC + ~CH. So what, if anything, could make it the case that we are talking about a (class) model in which CH is determinately true or false? There do not seem to be any causal relations between us and sets which could help us tie down reference (Putnam [1980], Field [1998b], [Martin 1976, 90–1]). One could always appeal to "reference magnets" à la Lewis [1983] and Sider

---

[8] Again, it would not help to add that "knowledge is the norm of belief." See, again, Chapter 4, fn. 23.

[9] See Korman and Locke [Forthcoming, § 4] for discussion of why the reliability challenge, or something like it, should grant the actual truth of the realist's beliefs and allow the realist to appeal to them in explaining their reliability.

[10] This would even be so if he merely granted that there were some atomic mathematical truths (without granting our particular mathematical theory), as Sharon Street sometimes seems to vis-à-vis evaluative truths in her [2006] work.

[11] Schechter [2010] distinguishes the task of explaining our coming to have a reliable mechanism for perceptual belief from the task of explaining how that mechanism works such that it is reliable.

[2011]. But that seems like theft over honest toil absent some account of what *makes* one model more natural than another, and of how naturalness facilitates reference. Accordingly, some philosophers hold that if a mathematical statement is undecidable with respect to the (first-order) axioms that we accept, then it is indeterminate.[12] But note that the fewer (determinate) truths one postulates, the fewer determinate truths one must explain reliability with respect to. Consequently, the reliability challenge arises *only to the extent that* the determinacy challenge can be answered.[13]

Of course, the challenge to explain the determinacy of our mathematical beliefs, realistically construed, remains. But it is surely no *harder* to meet in the moral case than in the mathematical. There are no undecidability results in the moral case to contend with. Moreover, interpreted at face value, moral truths are *about* concrete things, such as people, actions, and events, while mathematical truths are about the likes of numbers and (pure) sets. It is prima facie much less puzzling how predicates ascribed of the former could have determinate extensions.

The final point is that, despite Field's talk of objects, his formulation of the challenge does not really depend on an ontologically committal interpretation of the area, F.[14] That is, it does not target F-realism per se. What matters is that the area satisfies F-Aptness, F-Belief, F-Truth, and F-Independence, *and is significantly objective* (in the sense of Sections 1.2 and 1.6). It does not matter whether the area satisfies F-Face-Value. If Field's challenge *did* depend on an ontologically committal interpretation of the

---

[12] Even if every mathematical sentence had a determinate truth-value, it would not follow that we would have fixed on a determinate model. The Löwenheim–Skolem theorems ensure that any (countable) first-order consistent theory with an infinite model has a model of every other infinite cardinality. Moreover, not even the non-recursively enumerable theory, True Arithmetic, is ω-categorical (i.e., categorical in models of cardinality ω).

[13] The reliability challenge and the determinacy challenge are commonly conflated. See, e.g., Barton [2016]. Benacerraf's presentation was ambiguous between them, as it was between the challenge to explain the justification of our mathematical beliefs and the challenge to explain their reliability. Perhaps he believed that a causal answer was required in both cases. So, the apparent impossibility of establishing a causal connection between our "cognitive faculties and the objects known" would threaten our claim to knowledge and determinate reference. But, again, if we knew that determinate reference was impossible, then we would at most need to explain the reliability of our belief that so and so follows from such and such (and Benacerraf's discussion of "combinatorialism" suggests that we can do this [1973, 668]). (Actually, we would also need to explain the reliability of our belief that certain things do *not* follow from such and such, since we would presumably need to explain the reliability of our belief that such and such is consistent. This suggests that the view that our mathematical notions are maximally indeterminate is simply incoherent. A similar problem arises in connection with pluralist views. See Section 6.2.)

[14] See Hellman [1989] or Chihara [1990] for ontologically innocent interpretations of our mathematical theories. It is no wonder that Field takes the problem to depend on an ontologically committal interpretation of mathematics, since he himself appeals to primitive modal ideology of the sort described below (see, again, Section 3.5). See his [1980 and 1989, Introduction].

area, then it would have no application to moral realism. Although Mackie compares moral realism to belief in Plato's Forms (Mackie [1977, 28]), we saw in Section 1.5 that moral realism is not an ontological doctrine. It is an ideological doctrine in the sense of Quine [1951a]. And while some realists have tried to use this fact to show that there is no reliability challenge for moral realism (Scanlon [2014, 122]), this is a mistake. The problem is to explain the correlation between our beliefs and the truths. This is difficult because of the independence and the objectivity of the truths, not because of their ontology. We could even state the problem so as not to assume the existence of truths. The problem is to explain moral instances of the schema: if we believe that P, then P—where we *use*, and do not mention, "P."

Consider metaphysical possibility. It is frequently alleged that the reliability challenge arises equally for modal realism in the sense of Lewis [1986]. Again, Robert Stalnaker writes,

> It is a familiar objection to…modal realism that if it were true, then it would not be possible to know any of the facts about what is…possible.…This epistemological objection…may…parallel…Benacerraf's dilemma about mathematical…knowledge.   [1996, 39–40]

Now let us imagine that instead of postulating an infinity of concrete worlds, as David Lewis does, we take modal operators as primitive, on analogy with negation. So we accept that, say, it is metaphysically possible that there are aliens, even though there are no possible worlds, concrete or otherwise, and that truths about what is metaphysically possible are independent of human minds and languages. Suppose, also, that Lewis is systematically correct about the modal status of all propositions that he considers. When Lewis says that it is metaphysically necessary that P, it is, and when he says that it is contingent or impossible that P, it is. Lewis is simply wrong to think that such truths have anything to do with the existence of worlds, whether concrete or abstract. Given that there are no peculiarly modal entities with which to get in touch, do we avoid the mystery of how we are reliable with respect to truths of metaphysical modality? Of course not. We still need to know what explains the reliability of our beliefs in such propositions as that it is metaphysically possible that there are aliens, given that their truth is independent of human minds and languages, and is significantly objective. Likewise, even if the mind-and-language independent truth of such sentences as "giving to charity is good" do not owe anything to the existence of The Good, there is surely the question of what explains the reliability of our beliefs in such sentences—given that their truth is independent and significantly objective.

To sum up: Field's presentation of the reliability challenge, properly conceived, is quite powerful. It cannot be dismissed as a puzzle of no practical significance. It does not raise a general skeptical problem that has an analog in the perceptual case. It does not turn on doubts about the determinacy of reference. And it does not rely on an ontologically committal interpretation of the area. Nevertheless, it is unclear at a crucial juncture. It is unclear what it would take to *explain the reliability* of our beliefs of a kind, F, in the present sense. The key question is whether there is any sense of "explain the reliability" which satisfies these conditions:

Moral Unreliability: It appears impossible to explain the reliability of our moral beliefs, realistically construed.

Mathematical Reliability: It does not appear impossible to explain the reliability of our mathematical beliefs, realistically construed.

Undermining Inexplicability: If it appears impossible to explain the reliability of our beliefs of a kind, F, realistically construed, then this undermines them, so construed.

5.3   Causation, Explanation, and Connection

The most familiar interpretation of "explain the reliability" is suggested by Benacerraf. He writes,

> I favour a causal account of knowledge on which for X to know that S is true requires some causal relation to obtain between X and the referents of the names, predicates, and quantifiers of S. . . . [But] . . . combining *this* view of knowledge with the "standard" view of mathematical truth makes it difficult to see how mathematical knowledge is possible. . . . [T]he connection between the truth conditions for statements of number theory and any relevant events connected with the people who are supposed to have mathematical knowledge cannot be made out.   [1973, 671–3]

Let us interpret Benacerraf's suggestion as follows.

Answer 1 (Causation): In order to explain the reliability of our F-beliefs, it is necessary to show, for any one of them, that P, that there is a causal relation between our (token) belief that P and the subject matter of P.

Mathematical Reliability is evidently false under this interpretation. That was Benacerraf's point. But *Moral Unreliability* is false as well. Again, morality is *about*—that is, refers to or (first-order) quantifies over—the likes of people, actions, and events.[15] There *are* causal relations between such things and our moral beliefs. Finally, as was noted in Chapter 4, *Undermining Inexplicability* is false. The causal theory of knowledge, in anything resembling its original form, has been widely rejected for reasons that are independent of the reliability challenge (Field [2005, 77]). And yet, if it is implausible that *knowledge* that P requires a causal relation to obtain between our belief that P and P's subject matter (or the fact that P, or…), then it is even more implausible that *justified belief* that P requires the appearance that this is so.

However, even if Benacerraf's own interpretation of "explain the reliability" does not satisfy any of *Moral Unreliability, Mathematical Reliability,* and *Undermining Inexplicability*, there is a proposal in the neighborhood which might. It is one thing to require that the subject matter of our beliefs helps to *cause* those (token) beliefs, and it is another to require that their contents (or truth) help to *explain* those beliefs. To be sure, the present non-causal sense of "explain" is far from transparent, and one could reasonably doubt that there is any intelligible notion in the neighborhood. But supposing that there is something to the notion, we might worry that if we do not believe that P (non-causally) because it is true, then the truth of our belief that P would be coincidental. Geoffrey Sayre-McCord expresses the worry in the moral case as follows.

> The problem with moral theory is that moral principles…appear not to play a role in explaining our making the [judgments] we do. All the…work seems to be done by psychology, physiology, and physics.   [1988, 442]

Let us interpret the present proposal as follows.

Answer 2 (Explanation):   In order to explain the reliability of our F-beliefs, it is necessary to show, for any one of them, that P, that the fact that P helps to explain, even if not cause, our (token) belief that P.

---

[15] Again, one might think that it is also about properties, like generosity or goodness. But this is confused. See Section 1.5.

Insofar as the *Joyce–Street Thesis*[16] is plausible, so is *Moral Unreliability* under the present interpretation of "explain the reliability." And insofar as *Steiner's Thesis*[17] is plausible, it might be thought that *Mathematical Reliability* is plausible too. We saw in Section 3.5 that a fragment of arithmetic seems to be implied by every explanation whatever, and so certainly by any explanation of our (token) belief in any member of that fragment.

But, first, showing that, for any member of fragment of arithmetic truths, P, P helps to explain our belief that P is a far cry from showing that, *for any mathematical truth that we believe*, P, P helps to explain this. Second, showing that the content of our belief is *implied* by some explanation of it is not yet to show that that content helps to explain it in any useful sense of "explains," for reasons touched on Chapter 4.

Consider any recondite logical truth, P, that you believe. Then P is trivially a consequence of the explanation of your belief that P—since P is a consequence of every explanation. But the fact that P need not have had a role in explaining your belief that P. You may have come to believe that P by flipping a coin, or via lucky but erroneous computation. Indeed, had you believed ~P, it still would have been the case that P, and not ~P, was implied by any explanation of your coming to believe that ~P. There is a palpable sense in which it is just dumb luck that P is implied by any explanation of your coming to believe that P, for any such logical truth that you believe, P. Just as the *Quine–Harman Thesis**[18] requires more for empirical justification than indispensability, *Answer 2* requires more for explaining the reliability of our mathematical beliefs than showing that their contents are implied by some explanation of our coming to have them.

These considerations suggest the following weakening of *Answer 2*.

Answer 3 (Indispensability):   In order to explain the reliability of our F-beliefs it is necessary to show, for any one of them, that P, that P is implied by some explanation of our (token) belief that P, even if not in an explanatory way.

---

[16] Recall that the *Joyce–Street Thesis* says that it is not the case that the contents of any of our moral beliefs are implied by any explanation of our coming to have them, realistically construed. See, again, Section 4.1.

[17] Recall that *Steiner's Thesis* says that the contents of at least some of our mathematical beliefs, realistically construed, are implied by every explanation of our coming to have them. See, again, Section 4.1.

[18] Recall that the *Quine–Harman Thesis** says that our belief that P is empirically justified if and only if P plays an *explanatory role* in the best explanation of some of our observations. See, again, Section 3.8.

Answer 3 is just what we obtain when we frame Genealogical Debunking Arguments, discussed in Chapter 4, as instances of the reliability challenge. The challenge to explain the reliability of our moral beliefs is now assumed to require showing that their contents are implied by some explanation of them. But that means that, even if *Moral Unreliability* and *Mathematical Reliability* are both true, *Undermining Inexplicability* must fail for the reason that *Debunkers' Thesis* does.[19] Again, learning that the contents of our (token) beliefs of a kind, F, are not implied by any explanation of them gives us no reason to doubt that our F-beliefs have epistemically desirable qualities— such as sensitivity, safety, or objective probability. And while one can always gesture at a lack of connection between our beliefs and the truths, we saw in Section 4.5 that there seems to be no epistemological reason to *care* whether our beliefs exhibit such a connection—*given* that whether they do is independent of whether they are sensitive, safe, and objectively probable. In order to explain the reliability of our F-beliefs, realistically construed, it is *not necessary* to show that their contents are implied by some explanation of our having them.

5.4 Counterfactual Dependence

Field is under no illusions on this point. When discussing his challenge as it applies to logic, he explicitly rejects *Answer 3* on the ground that such an explanation is not *unified* [Field 1996, 372 n. 13].[20] What does it take for an explanation to be unified? Field writes,

> The idea of an explanation failing to be "unified" is less than crystal clear, but another way to express what is unsatisfactory about [an non-unified explanation] is that it isn't *counterfactually persistent*...it gives no sense to the idea that if the...facts had been different then our...beliefs would have been different too.   [1996, 371, italics in original]

[19] Recall that *Debunkers' Thesis* says that if it appears that it would be a coincidence if our beliefs of a kind are true, realistically construed, then this undermines them, so construed (where the sense of "coincidence" at issue must satisfy *Moral Debunking* and *Mathematical Vindication*). See, again, Section 4.2.

[20] His argument actually bears on the sufficiency of showing that the contents of our F-beliefs are implied by some explanation of our coming to have them. But Field evidently rejects the necessity claim, since he holds that mathematical pluralism, to be discussed, affords an answer to the reliability challenge, and this view, by itself, does not imply that the contents of our mathematical beliefs are implied by some explanation of those beliefs—or, indeed, by some explanation of any concrete event. See Section 5.8.

Or again,

> The Benacerraf problem...seems to arise from the thought that we would
> have had exactly the same mathematical...beliefs even if the mathematical...
> truths were different...and this undermines those beliefs.   [2005, 81][21]

What does Field mean by "different"? He apparently means arbitrarily
different. He is particularly concerned by the fact that "we can assume,
with at least some degree of clarity, a world without mathematical
objects..."[2005, 80–1].

The reliability challenge is often understood similarly in the moral case.
Matt Bedke writes,

> Whatever form the moral facts or properties take, one would have the
> very same moral...beliefs because such things are causally determined,
> and the causal order has not changed.   [2009, 196]

Let us, therefore, interpret Field's suggestion as follows:

Answer 4 (Counterfactual Persistence):   In order to explain the reliability of
our F-beliefs, it is necessary to show, for any one of their contents, that P,
that had the F-truths been arbitrarily different such that ~P, we would not
still have believed that P.

Answer 4 is technically incomplete. Modal conditions on knowledge and—
we may add—justified belief require relativization to methods of belief
formation. That our F-belief that P would have been false had the F-truths
been different only because the closest worlds in which the F-truths are different are worlds in which we decide what to believe by flipping a coin is not
undermining. Let us, therefore, reformulate *Answer 4* as follows:

Answer 4 (Counterfactual Persistence): In order to explain the reliability of
our F-beliefs, it is necessary to show, for any one of their contents, that P,
that had the F-truths been arbitrarily different such that ~P, we would not
still have believed that P (had we used the method that we actually used to
form them).

---

[21] See also Field's discussion in his [1990] work.

Whether *Moral Unreliability* and *Mathematical Reliability* are true under this interpretation of "explain the reliability" is immaterial. *Undermining Inexplicability* is still false. As we saw in Section 4.3, had the perceptual truths—the truths of ordinary perception—been *arbitrarily* different, our perceptual beliefs may well have been the same. In particular, had the atomic such truths been systematically false, because we were brains in vats, we would still have believed that, for example, we had hands. What we can hope to show is that our perceptual beliefs are *sensitive* (Nozick [1981, ch. 3]). That is, we can hope to show that had the content of any one of our perceptual beliefs been false, we would not still have believed it (using the method that we actually used to form that belief). This suggests the following revision to *Answer 4*:

Answer 5 (Sensitivity): In order to explain the reliability of our F-beliefs it is necessary to show that our F-beliefs are *sensitive*—that is, that for any one of them, that P, had it been that ~P, we would not still have believed that P (using the method that we actually used to determine whether P).[22]

Rather than requiring that we do not believe P in ~P worlds in which the F-truths are varied in any which way, *Answer 5* just requires that we do not believe P in the *closest* ~P worlds (in which we still determine whether P using the method that we actually used). Whether *Moral Unreliability*, *Mathematical Reliability*, and *Undermining Inexplicability* are true under this interpretation of "explain the reliability" requires discussion. There is an influential argument for *Moral Unreliability* and *Mathematical Reliability* which implicitly assumes *Undermining Inexplicability*.

5.5 Selection for Truth

It is widely alleged that there is an evolutionary argument for the sensitivity of our mathematical beliefs. At first approximation: it would have benefited our ancestors to have true, rather than false, mathematical beliefs! Hence, assuming that the requisite cognitive mechanisms are heritable, evolution would have selected for them. The following quotations are representative.

---

[22] Again, "for any one of them" is likely too strong a quantifier. Perhaps "for a typical one of them," or "for most of them," or "for a weighted majority of them" (relative to some weighting) would be better. I ignore this complication.

The rich consilience of arithmetic views is no sheer fluke, we can be sure.... What kind of explanation of the consilience would fail to be an explanation of the tendency to get the matter right?...The right explanation presumably involves capacities that were reproduction-enhancing among our ancestors, and a history of invention and correction that these capacities, in extended application, made possible.   [Gibbard 2003, 257]

Any reasonable explanation for why it was to our ancestors' reproductive advantage to have a hardwired belief that 1 + 1 = 2 (say) will depend on that beliefs being *true*: a false arithmetic belief just isn't going to be useful. [Joyce 2008, 217, italics in original]

Humans...appear to have an innate sense of number, which can be explained by the advantage of reasoning about numerosity during our evolutionary history. (For example, if three bears go into the cave and two come out, is it safe to enter?) But the mere fact that a number faculty evolved does not mean that numbers are hallucinations.... [T]he number sense evolved to grasp abstract truths in the world that exist independently of the minds that grasp them.[23] [Pinker 2002, 192]

By contrast, the challenge to show that our moral beliefs are sensitive is supposed not to admit of a similar answer. For example, Walter Sinnott-Armstrong writes,

The evolutionary explanations [of our having the moral beliefs that we have] work even if there are no moral facts at all. The same point could *not* be made about mathematical beliefs. People evolved to believe that 2 + 3 = 5, because they would not have survived if they had believed that 2 + 3 = 4, but the reason why they would not have survived then is that it is *true* that 2 + 3 = 5. [2006, 46, italics in original]

Roger Crisp writes,

In the case of mathematics, what is central is the contrast between practices or beliefs which develop because that is the way things are, and those that do not. The calculating rules developed as they did because [they] reflect mathematical truth. The functions of...morality, however, are to be

---

[23] Unlike the other authors quoted here, Pinker is apparently open to a similar account of moral judgment in his [2002] work.

understood in terms of well-being, and there seems no reason to think that had human nature involved, say, different motivations then different practices would not have emerged. [2006, 17]

And Street claims:

[M]aking [moral] judgements contributed to reproductive success not because they were true or false, but rather because they got our ancestors to respond to their circumstances with behavior that itself promoted reproductive success in fairly obvious ways: as a general matter, it clearly tends to promote reproductive success to do what would promote one's survival, or to accord one's kin special treatment, or to shun those who would harm one. [2006, 128–9]

Street appears to draw a false dichotomy. She suggests that either we were selected to have the moral (or, more generally, evaluative) beliefs that we do have because they were true, or that we were selected to have them because they contributed to reproductive success. However, if we were selected to have the moral beliefs that we do have, then we were certainly selected to have them because they contributed to reproductive success! That is basically just what "selected to have" means. The question, as Sinnott-Armstrong emphasizes, is whether it contributed to reproductive success to have the moral beliefs that we do have because they were true.[24]

But what does this question really come to? The key issue is whether, if P is some moral truth that selective forces led us to believe, then, had it been that ~P, it would have benefited our ancestors to believe that ~P.[25] Let us say that we were selected to have *true moral beliefs per se* if this counterfactual holds, and that we were selected to have *moral beliefs with property, G, which are, in fact, true* if, had it been that ~P, but beliefs in P still had property, G, then it still would have benefited our ancestors to believe that P.[26] Then note that only the thesis that we were selected to have true F-beliefs per se could explain the reliability of our F-beliefs in the present sense. If we were merely selected to have F-beliefs with property G which are, in fact, true, then, if the closest worlds in which our F-belief that P is false are still worlds in

---

[24] In other words, the question is whether there was selection *for* true moral beliefs, as opposed to selection *of* them, in the sense of Sober [1984, 98–101].

[25] I am not putting forth this counterfactual as a conceptual analysis of selection claims. I am saying that it is what matters for the argument that our moral beliefs are not sensitive.

[26] See Field [2005] for roughly this distinction. See, again, Sober [1984] for related discussion.

which F-beliefs have property G, then we may still have believed that P (using the method that we actually used to determine whether P). We were selected to have F-beliefs with property, G, which are, in fact, true *whenever* we were selected to have the F-beliefs that we do have, and those beliefs happen to be true—even if only thanks to a "coincidence." Again, it must be granted in this dialectical context that our moral beliefs are actually true. But this is not to grant that they are sensitive.

Of course, the thesis that we were literally selected to have the moral or mathematical beliefs that we do have, realistically construed, whether because they were true or not, is almost certainly too simple.[27] Moral and mathematical beliefs, realistically construed, seem to have too recent an origin to have been selected for. It could be that the only credible view in the neighborhood is that we were selected to have cognitive mechanisms which involve dispositions to form belief-like representations with contents that are somehow related to those of our moral or mathematical beliefs, realistically construed (Butterworth [1999], Dahaene [1997], De Cruz [2006], Joyce [2007] Pantsar [2014], Relaford-Doyle and Núñez [2018]).[28] But the differences between the simple and credible views will be unimportant here. So, I will simply assume that we were literally selected to have at least some of the moral and mathematical beliefs that we do have, realistically construed. The question remains whether we would thereby have been selected to have true moral or mathematical beliefs per se.

5.6   The Mathematical Indifference of Evolution

Why is it so widely supposed that we would have been selected to have true mathematical beliefs per se? Joyce summarizes the reason as follows:

> There is…evidence that the distinct genealogy of [arithmetic] beliefs can be pushed right back into evolutionary history: that natural selection has provided humans with an inbuilt faculty for simple arithmetic

---

[27] This is something that Street herself emphasizes in in her [2006] work.

[28] In the mathematical case, Pantsar claims that "there are important connections between the ANS [which allows us to non-inferentially judge numerosities – not numbers! – in our field of vision, and keep track of them in working memory], the symbolic presentation of numerosities, and counting. There is starting to be way too much correlation to be explained away simply as a coincidence. The data clearly points to the direction that our verbal ability to deal with numerosities was built to accommodate the primitive non-verbal [ANS] system" [Pantsar 2014, 4209].

(Butterworth 1999)... [But] we have no grasp of how [belief that $1 + 1 = 2$] might have enhanced reproductive fitness independent of assuming its truth. False mathematical beliefs just aren't going to be very useful.... The truth of '$1 + 1 = 2$' is a background assumption to any reasonable hypothesis of how this belief might have come to be innate.   [2007, 182]

We can illustrate Joyce's reasoning with the following example:

Imagine that there are two lions in the meadow, one behind bush A and the other behind bush B. Two of our ancestors, Jenn and Joe, are hiding behind a tree, aware of the lions, and gawking at a bush of nutritious berries that is growing there. Jenn believes that the one lion and another lion make two lions in all, while Joe believes that one lion and another lion make zero lions in all. Then Jenn will have a reproductive advantage over Joe. Joe will be more likely than Jenn ceteris paribus to walk out into the meadow and get eaten by two lions, and, so, less likely to pass on his genes. And any explanation of this will imply that one lion and another lion really do make two, and not zero, lions in all.

There is more than one problem with this example.[29] But the immediate problem should be familiar from Section 4.3. The example seeks to establish the wrong conclusion.[30] Joyce intends to show that we must presuppose the contents of our arithmetic beliefs in any evolutionary explanation of our coming to have them. This indispensability point does nothing to establish the counterfactually loaded thesis that we were selected to have true arithmetic beliefs per se. If it did, then, again, for any logical truth that we were selected to believe, P, it would be *trivial* to show that we were selected to have a true belief in P per se. As a logical truth, P is implied by every explanation. So, it is certainly implied by any evolutionary explanation of our coming to believe that P. But, even if we were selected to have true logical beliefs per se, this is not trivial.

In order to argue that we would have been selected to have true mathematical beliefs per se, we need to argue for a counterfactual. We need to argue that *had $1 + 1 \neq 2$* (or, what comes to the same thing, *had our belief*

---

[29] I discuss the other problem below.
[30] This mistake is very common. See Street [2006, 160–1, n. 35] and Griffiths and Wilkins [2015, § 4.5] for other examples.

that 1 + 1 = 2 been false), it would have benefited our ancestors to believe that 1 + 1 ≠ 2. In terms of the above example, we might argue as follows.

Imagine that one lion and another lion *really did* make zero lions in all. Then Jenn, who believes that one lion and another lion make two lions in all, would no longer have a reproductive advantage over Joe, who believes that one lion and another lion make zero lions in all. In particular, Joe would no longer be more likely, ceteris paribus, to walk into the meadow and get eaten by two lions. There would not be any lions behind the bushes, so Joe could not be eaten by any. Moreover, Jenn would forgo nutritious berries in the meadow on account of her false belief. So, had the arithmetic truths been different, it would have benefited our ancestors to have correspondingly different arithmetic beliefs.

Let us assume for the sake of argument that the counterfactual "Imagine that one lion and another lion really did make zero lions in all" is intelligible. (If it is not, then one cannot argue that, had the arithmetic truths been different, it would have benefited our ancestors to have correspondingly different arithmetic beliefs a fortiori. One could, at most, argue that there is no intelligible question of whether this is so.[31]) Nevertheless, the argument on behalf of Joyce et al. still fails. It trades on an equivocation between mathematical truths, realistically construed, and (first-order) logical truths. Suppose that what is being imagined is that if there is "exactly one" lion behind bush A and "exactly one" lion behind bush B, and no lion behind bush A is a lion behind bush B, then there are no lions behind bush A or B (where the numerical quantifiers here are definable in terms of ordinary quantifiers plus identity). That is, suppose:

$$[\exists x(Lx \ \& \ Ax \ \& \ \forall y([(Ly \ \& \ Ay) \rightarrow (x = y)])) \ \& \ \exists x(Lx \ \& \ Bx \ \& \ \forall y([(Ly \ \& \ By) \rightarrow (x = y)]) \ \& \ {\sim}\exists x(Lx \ \& \ Ax \ \& \ Bx)] \rightarrow {\sim}\exists x(Lx \ \& \ (Ax \ v \ Bx))$$

where "Lx" means that x is a lion, "Ax" means that x is behind bush A, and "Bx" means that x is behind bush B.

Then it may be true that Joe would not be more likely to get eaten than Jenn. Again, there would not be any lions in the meadow, so Joe could not be

---

[31] It might be thought that such an argument would itself establish a lack of epistemic parity between the moral and mathematical cases. But this is incorrect. See § IV of my [2012] work.

eaten by any. But to imagine the proposition expressed by the sentence above is not to imagine that $1 + 1 = 0$, realistically construed. It is to imagine a bizarre variation on the *(first-order) logical truth* that if there is "exactly one" lion behind bush A and "exactly one" lion behind bush B, and no lion behind bush A is a lion behind bush B, then there are "exactly two" lions behind bush A or B (where "exactly one" and "exactly two" are abbreviations for constructions out of ordinary quantifiers and identity, as above).

Realistically construed, the claim that $1 + 1 = 0$ speaks of *numbers*. It says that the number 1 bears the plus relation to itself and to 0. What we really need to ask is what would happen if the number 1 bore the plus relation to itself and to 0 *and the (first-order) logical truth* that if there is exactly one lion behind bush A, and exactly one lion behind bush B, and no lion behind bush A is a lion behind bush B, then there are exactly two lions behind bush A or B, *held fixed*. Then, if Joe's belief that $1 + 1 = 0$ would have had any effect on his reproductive fitness at all, it seems that Joe *would* still be more likely to get eaten than Jenn. (If Joe's belief that $1 + 1 = 0$ would *not* have had any effect on his reproductive fitness, then we could not use the above example to show that we were selected to have true mathematical beliefs per se.) There would be two lions behind bush A or bush B (in the first-order quantificational sense of "two"), and Joe would be disposed to behave as if there were no lions there. For example, he might walk out from behind the tree, rather than staying hidden behind it for fear of being eaten. Assuming that the *(first-order) logical truths* held fixed, it seems that Joe would have been at a disadvantage after all.

The point can be stated more intuitively. If our ancestors who believed that $1 + 1 = 2$ had a reproductive advantage over our ancestors who believed that $1 + 1 = 0$, the "reason" (to use Sinnott-Armstrong's language) is that corresponding (first-order) logical truths obtained. Or, to use Street's language, this is "because" such logical truths obtained. (But note that, contrary to what these authors suggest, the pertinent sense of "reason" and "because" *has nothing to do with explanatory indispensability!*) Jenn, who believed that $1 + 1 = 2$, had a reproductive advantage over Joe, who believed that $1 + 1 = 0$, in the above scenario "because" if there is exactly one lion behind bush A, and there is exactly one lion behind bush B, and no lion behind bush A is a lion behind bush B, then there are exactly two lions behind bush A or B. In other words, Jenn did *not* have an advantage over Joe because Jenn's belief that $1 + 1 = 2$ was true. She had an advantage over Joe because her belief appropriately aligned with logical truths about her surroundings.

What is the upshot? It is *not* that we were selected to have true (first-order) logical beliefs, rather than true mathematical beliefs per se. I have taken no stand on the case of logic.[32] The upshot is that logical truths would *explain the benefit* conferred on creatures who had the arithmetic beliefs that we do.

It might be objected that this strategy for explaining the usefulness of true arithmetic beliefs in terms of (first-order) logical truths is not sufficiently general [Braddock, Mogensen, and Sinnott-Armstrong 2012]. After all, (first-order) logical truths like the one above are about individual objects, such as lions or cliffs. But in order to explain the general fact that creatures who believed that $1 + 1 = 2$ had an advantage over those who did not, it seems insufficient to cite a fact about individual lions, for example. The fact that if there is exactly one lion behind bush A, and there is exactly one behind bush B, and no lion behind bush A is behind bush B, then there are exactly two lions behind bush A or B does not explain the general fact in question.

The point is fair so far as it goes, but it does not undercut the approach. We could always ascend to second-order logic. Creatures who believed that $1 + 1 = 2$ had an advantage over creatures who believed that $1 + 1 = 0$ because, for any properties F, G, and H, if there is exactly one F that is G, and there is exactly one F that is H, and no G is an H, then and there are exactly two Fs that are G or H. But reliable belief about second-order consequence seems to be comparably mysterious as reliable belief about natural numbers (even assuming, contra Quine, that second-order logic is *not* "set theory in sheep's clothing" [Quine 1986, ch. 5]).[33] It is better to appeal to the first-order schemas one obtains by dropping off just the second-order quantifiers from second-order logical truths like the ones mentioned. The resulting explanations explain in exactly the sense that any first-order theory with a recursive, but infinite, set of axioms does.[34]

---

[32] The case of logic really factors into two cases. One question is: had the truths about *what follows from what* been different, would it have benefited our ancestors to have correspondingly different beliefs? The answer seems to be that, under the dubious assumption that they had (proto-)beliefs about such things at all, their beliefs would not have been so different. Had, e.g., Disjunctive Syllogism not been valid, our ancestors still would have (proto-)believed that it was, given (dubiously!) that they believed this to start with. But this technically shows that we would not have been selected to have true *metalogical*, not logical, beliefs. The logical case requires that we vary the *logical truths themselves*—not just their validity. Had, e.g., it been false that either there are lions behind the bush or it is not the case that there are lions behind the bush, would it still have benefited our ancestors to believe this disjunction (if they did)? There may be no determinate answer. See Clarke-Doane [2012, § III].

[33] This is because the reliability challenge for an area does not depend on whether that area is about special entities, like sets. See, again, Section 5.2.

[34] There remains the question of how to handle the case of geometry and the case of arithmetic truths with no (first-order) surrogate logical truths. See my [2012, § 3] for more on this.

5.7 The Inadequacy of the Sensitivity Challenge

So, the evolutionary argument that even our most rudimentary arithmetic beliefs are sensitive does not stand up to scrutiny. If *Mathematical Reliability* is true, this is not because we were "selected to have true mathematical beliefs." But if *Mathematical Reliability* is false, then it might be thought that *Moral Unreliability* is true a fortiori. On the contrary, an analogous argument in the moral case may actually *succeed*—at least for atomic truths (notwithstanding the quotation from Bedke in Section 5.3, and the quotations from Joyce and Michael Ruse in Section 4.3). Suppose, as in the mathematical case, that we were selected to have at least some of the moral beliefs that we do have. Perhaps, for example, we were selected to believe that feeding our offspring when they are hungry is generally good. Moreover, assume, as in the mathematical case, that this is true.[35] What would our ancestors have believed, had feeding our offspring when they are hungry *not* been generally good? Had it not been generally good to feed their offspring when they were hungry, the world would have been different in non-moral respects— since the closest worlds in which this is so are still worlds in which the explanatorily basic moral truths which fix the supervenience of the moral on the non-moral are the same. Perhaps our ancestors' offspring would have been hungry more often than is good for them. Or maybe they would have been able to feed themselves, and our feeding them might have upset their development. Or perhaps they would have broken out into a life-threatening rash if they were fed by their parents, rather than by other relations. Had any of these (unlikely!) events transpired, however, our ancestors' moral beliefs may well have varied correspondingly. Had, for instance, our ancestors' offspring broken out into a life-threatening rash whenever our ancestors, as opposed to other relations, fed them, presumably our ancestors would not have (proto-)believed that it was good to feed their offspring. The upshot is that, in contrast to the mathematical case, there is no obvious obstacle to arguing that we would have been selected to have true atomic moral beliefs per se.[36]

---

[35] Again, the latter assumption is legitimate, since the apparent impossibility of showing that our moral beliefs are sensitive is supposed to *undermine* them. So, it gives us reason to doubt that if they are true, then they are sensitive, if it gives us reason to doubt that they are sensitive. See Korman and Locke [Forthcoming, § 4], which, coincidentally, uses the same example of a rudimentary moral proposition.

[36] Depending on how we measure closeness of worlds, it *might* be possible to argue that we were selected to have true *impure* elementary arithmetic beliefs per se. Had the number of

Why in the world would it have benefited our ancestors to "grasp the independent [moral] truth" per se [Street 2006, 130]? Because selection for true belief per se *just is* the kind of counterfactual co-variation described. It has nothing to do with whether we can "explain why human beings tend to make the [moral] judgments that they we do... [without] suppos[ing] that these... are *true*" [Street 2008: 208–9, italics in original]. Of course, we could just decide to *use* the term "selection for true belief per se" so that it did. But, in that case, whether we were selected to have true F-beliefs per se would have nothing to do with whether our F-beliefs are sensitive, or even safe (see Sections 4.3 and 4.4). The reason that the argument for sensitivity has force in the moral, but not mathematical, case is that atomic moral truths ascribe metaphysically supervenient properties to concrete things in the world around us, like people. Had those things exemplified different moral properties, they would have exemplified different causally efficacious ones—and our moral beliefs may have reflected the difference. By contrast, mathematical truths ascribe (vacuously supervenient) properties to the likes of numbers. Such things have no causally efficacious properties. So they have no such properties to which we can respond.

The reasoning above just recapitulates the reasoning used in Section 4.3 to show that the Joyce–Street Thesis gives us no reason to doubt the sensitivity of our atomic moral beliefs. Indeed, we could have bypassed evolutionary speculation altogether, and argued directly for their sensitivity. Again, for any atomic moral truth that we believe, A is M, had A not been M, then A would have been different in non-moral respects, and had A been different in non-moral respects, then our moral beliefs about A would have reflected the difference. Had, for example, the people from Harman's example (in Section 3.6) been petting the cat rather than pouring gasoline on it and igniting it, we would not have believed that they were doing something wrong to the cat (using the method that we actually used). Such an argument would do nothing to *dialectically justify* our moral beliefs, realistically construed, since it assumes that they are actually true. But I am not trying to dialectically justify them. I am saying that, assuming that they are true, we can explain the reliability of our atomic ones, in the sense of showing that they are sensitive.

---

lions in the meadow not been equal to 2, perhaps our ancestors would not have believed that it was (where "the number of lions in the meadow" is a singular term picking out a number). But, even if this is so, it would not show that we were selected to have true *pure* arithmetic beliefs per se, which is what was at issue in Section 5.5 (see, again, Section 1.1 on the individuation of areas). Moreover, there are reasons to think that it is not so. See my [2019, Section 7].

However, while we would seem to be able to show that our atomic moral beliefs are sensitive, realistically construed, the challenge to show that our beliefs of a kind are sensitive turns out to be too demanding, in general. That is, *Undermining Inexplicability* is false under the present interpretation of "explain the reliability."[37] The fact that our mathematical beliefs are insensitive is a special case of a general one. *For pretty much any metaphysically necessary truth, P,* our belief that P is insensitive, even if true. (The qualifier "pretty much" is needed, since who knows what we would have believed had, say, our belief that if there are dogs, then there are dogs been false.) This applies, in particular, to *explanatorily basic* moral truths which fix the conditions under which a moral property is instantiated, as a matter of metaphysical necessity. Indeed, for nearly any bridge law linking metaphysically subvenient to supervenient properties, it seems that had that law been false, our beliefs would have been the same (had we used the method that we used to determine whether it was true). For instance, had particles arranged "paper-wise" failed to compose a piece of paper, it seems that we still would have believed that they did.

It might be thought that we could just be skeptics about all metaphysically necessary truths, realistically construed. Again, Casullo suggests in one of the quotations that began this chapter that the reliability challenge threatens belief in all such truths. But skepticism about necessary truths is hard to contain. If I believe that I am looking at a piece of paper, but my belief in the metaphysically necessary bridge law that particles arranged paper-wise compose a piece of paper is undermined, then it is hard to see how my belief that I am looking at a piece of paper could fail to be. This is not incontrovertible for the reason that it is not incontrovertible that if I know that I have hands, then I know that there is an external world. I have assumed a closure principle, which advocates of a sensitivity condition on knowledge may deny (Nozick [1981, 227–9]). But it is difficult to see how our beliefs about the conditions under which properties are instantiated could be rationally insulated from our beliefs ascribing them.

---

[37] It is also not demanding enough. The argument that our moral beliefs are sensitive works for *any supervenient properties* of concrete things, even if epiphenomenal. It even works for aesthetic or astrological properties, given that these would be metaphysically supervenient. For any such property, F, we can *trivially* show that our *atomic* F-beliefs are sensitive, using the schema above. The challenge to show that our beliefs are sensitive only appears impossible to meet when the truths are *about*—i.e., (first-order) quantify over—abstract things, like numbers, or when they float free of ontology altogether, as would primitive modal truths (though there are conceivable exceptions to this rule). While there is a pressing reliability challenge for realism about such truths, it arises a fortiori for moral, aesthetic, or astrological realism.

To be clear: I am *not* arguing that the sensitivity challenge fails because mathematical or explanatorily basic moral truths are necessary—so our corresponding beliefs are vacuously sensitive. The sense in which mathematical and explanatorily basic moral truths have any claim to being necessary is not a maximal sense of "necessary" (see, again, Section 3.4). So, *even assuming a standard semantics for counterfactuals*, our corresponding beliefs are not vacuously sensitive. Of course, if the mathematical and explanatorily basic moral truths *were* maximally necessary, or if the standard semantics turned out to be untenable, then the sensitivity challenge would fail a fortiori.[38]

5.8 Contingency

I have argued that none of *Answers 1–5* satisfy all of the conditions, *Moral Unreliability*, *Mathematical Reliability*, and *Undermining Inexplicability*. So, *if* there is any sense of "explain the reliability" such that (a) it appears impossible to explain the reliability of our moral beliefs, realistically construed, (b) it does not appear impossible to explain the reliability of our mathematical beliefs, so construed, and (c) the apparent impossibility of explaining the reliability of our beliefs undermines them, so construed, *then* that sense has nothing to do with the challenge to establish a causal, explanatory, logical, or even counterfactual dependence between our beliefs and the truths. Not only do none of *Answers 1–5* seem to satisfy both *Moral Unreliability* and *Mathematical Reliability*; none seems to satisfy *Undermining Inexplicability*.

However, there are two ways to have false beliefs of a kind, F. First, it could happen that the F-truths are different while our F-beliefs fail to be correspondingly different. Second, it could happen that our F-beliefs are different while the F- truths fail to be correspondingly different. Even if the first possibility is inapt when the truths are supervenient or necessary, the latter remains. Perhaps there is an interpretation of "explain the reliability" in terms of the latter such that *Moral Unreliability*, *Mathematical Reliability*, and *Undermining Inexplicability* are all true.

Despite Field's remarks on counterfactual persistence, he seems to suggest that Benacerraf's challenge stems from the possible variation of our beliefs, not the truths, when discussing so-called "mathematical pluralism"

---

[38] See Williamson [Forthcoming] for a defense of the standard semantics for counterfactuals.

(discussed briefly in Sections 1.6 and 4.4, and in detail in Chapter 6). For example, he writes,

> [Mathematical pluralists] solve the [reliability challenge] by articulating views on which though mathematical objects are mind independent, any view we had had of them would have been correct.... [2005, 78]

Keeping in mind the need to relativize to methods of belief formation, let us interpret Field's proposal as follows:

Answer 6 (Fail-safety): In order to explain the reliability of our F-beliefs it is necessary to show, for any one our F-beliefs, that P, that had we believed that ~P, our belief still would have been true (had we used the method that we actually used to form them).

But, even bracketing *Moral Unreliability* and *Mathematical Reliability*, *Undermining Inexplicability* is clearly false under this interpretation. Perhaps the closest world in which we believe the negation of, for example, the Axiom of Foundation is a world in which our mathematical beliefs are inconsistent. Nor would it help to weaken *Answer 6* as follows.

Answer 7 (Consistent Fail-safety): In order to explain the reliability of our F-beliefs it is necessary to show, for any one of their contents, that P, that had we believed that ~P, and our F-beliefs remained consistent, then our belief still would have been true (had we used the method that we actually used to form them).

Undermining Inexplicability remains false under this interpretation because, had we believed that there were ghosts but our perceptual beliefs remained consistent, then our belief that, for example, we are seeing one would have been false too. The closest worlds in which we have such a belief are worlds in which we are deluded. *Answer 6* and *Answer 7* falter because they equate a condition that is constitutive of realism—namely, F-Independence from Section 1.2—with an epistemological problem. To say that had we believed ~P our belief that P would have been false (if it is actually true) is just to say that the truth that P does not counterfactually depend on our belief. As I mentioned in Section 4.4, in order to turn this condition into a problem we must add that *we could have easily believed that ~P*. It then follows that we could have easily had a *false* belief as to whether P (assuming that the truth that P could not have easily been different).

Keeping in mind the need to relativize to methods of belief formation, this suggests a final interpretation of "explain the reliability."

Answer 8 (Safety): In order to explain the reliability of our F-beliefs it is necessary to show that they are *safe*—that is, for any one of them, that P, that we could not have *easily* had a false belief as to whether Q, where Q is any proposition similar enough to P (using the method that we actually used to determine whether P).[39]

Answer 8 is the only answer to the question with which we began of which I am aware that has some claim to satisfying *Undermining Inexplicability*—although it is a far cry from the glosses offered by Benacerraf and Field. As illustrated in Section 4.4, evidence that we could have easily had a false belief as to whether P (using the method that we actually used to form our belief) is at least often an undermining (as opposed to rebutting) defeater of our belief that P. It gives us reason to give up our belief that P, but not by giving us direct reason to believe that ~P.

Moral Unreliability is plausible because our moral beliefs appear to be highly contingent. Recall the kind of moral disagreement that was surveyed in Chapter 2. The problem with such disagreement is not, contra Mackie [1977, 26], that it is best explained by the thesis that there are no atomic moral truths, realistically construed. Moral realism has no implications at all for the distribution of opinion—at least if moral facts would be causally inert. So, the existence and character of moral disagreement is no evidence for moral anti-realism (or realism). The problem with such disagreement is that it persists among "peers" who seem to share important biographical details with us. Such peers may have the same non-moral evidence, may share our standards of argument, may be as intelligent, educated, sincere, attentive, and so on. The main characteristics that may vary between us are highly contingent sensibilities. Had we grown up in a different town, read different books, or had different mentors—scenarios that could have "easily" obtained—our moral outlook might have been very different. Whether knowledge of this is undermining by itself, it at least threatens to be *qua* evidence that we could have easily had false moral beliefs (Hirvela, [2017], Mogensen [2016], White [2010]). As Brian Leiter notes,

---

[39] For the need to complicate the definition of safety so as to concern similar enough beliefs, see Section 4.6.

[We need not] exploit anthropological reports about the moral views of exotic cultures.... [There is] disagreement at the heart of the most sophisticated moral philosophies of the West, among philosophers who very often share lots of beliefs and practices and who, especially, in the last century, often share the same judgments about concrete cases.... [T]hese philosophers remain locked in apparently intractable disagreement about the most important, foundational issues about morality.   [2010]

G. A. Cohen makes the connection between disagreement and contingency explicit. He writes,

I... [had] a strongly political upbringing.... [I]ntense belief [was] induced in propositions that other people regard as false; indeed, very often... obviously false.... It should give us pause that we would not have beliefs that are central to our lives—beliefs, for example, about important matters of politics and religion—if we had not been brought up as we in fact were. It is an *accident of birth and upbringing* that we have them, rather than beliefs sharply rival to them.   [2000, 9, italics added]

(Cohen is officially concerned here with disagreement over "politics and religion." But his worry evidently arises equally in the case of morality.)

What about *Mathematical Reliability*? A mathematical realist might appeal to evolution to defend it. Even if we were not selected to have true mathematical beliefs per se, perhaps we were selected to have *the mathematical beliefs that we do have, which are, in fact, true*. Indeed, perhaps this is what Gibbard really means in claiming that "The rich consilience of arithmetic views is no sheer fluke, we can be sure...." If we could argue that, for any mathematical proposition that we believe, P, we were evolutionarily disposed to believe that P, then, we could argue that our mathematical beliefs are safe in two steps. First, the mathematical truths could not have easily been different, because they are metaphysically necessary. Second, belief in P—or, better, belief in something from which we are disposed to infer P—was evolutionarily inevitable. Hence, for any such P, we could not have easily believed ~P. So, we could not have easily had a false belief as to whether P.[40] It still does not quite follow that our belief that P is safe. We must add that there is no Q that is similar enough to P such that we could have easily had a false belief as to whether Q. But, if this could be argued, then it would follow that our mathematical beliefs are safe.

[40] See Clarke-Doane [2016b, § 2.3] for a slightly more careful version of such an argument.

But such an argument is hard to take seriously. Even in the case of arithmetic, "the development from...proto-arithmetical experiences...to our axiomatic systems of arithmetic was hardly inevitable" [Pantsar 2014, 4219].[41] Moreover, however inevitable our arithmetic beliefs are, the arithmetic truths dramatically underdetermine the content of our mathematical theories. For all manner of non-arithmetic mathematical propositions that we believe, that P, it seems that we could have easily believed that ~P (using the method that we actually used to determine whether P). As we saw in Chapter 2, our mathematical beliefs also "seem to be greatly influenced by [our] training and...environment" [Cohen 1971, 10].[42] What is more, the character of mathematical disagreement suggests that our training and environment *could have easily been different* in the relevant sense. Again, John Bell and Geoffrey Hellman write,

Contrary to the popular (mis)conception of mathematics as a cut-and-dried body of universally agreed upon truths...as soon as one examines the foundations of mathematics [the question of what axioms are true] one encounters divergences of viewpoint...that can easily remind one of religious, schismatic controversy.   [2006, 64]

Pavel Pudlák makes the connection to contingency explicit. He writes,

Imagine that the Axiom of Determinacy [which is inconsistent with the Axiom of Choice] had been introduced first, and before the Axiom of Choice was stated the nice consequences of determinacy, such as the measurability of all sets, had been proved. Imagine that then someone would come up with the Axiom of Choice and the paradoxical consequences were proved. Wouldn't the situation now be reversed in the sense that the Axiom of Determinacy would be 'the true axiom', while the Axiom of Choice would be just a bizarre alternative?   [2013, 221][43]

---

[41] Indeed, Relaford-Doyle and Núñez [2018] argue that not even the principle that every natural number has a successor is innate, or even universal *today* among educated non-mathematicians.

[42] Recall that the quotation is from *Paul* Cohen, the mathematician, not G. A. Cohen.

[43] Joel David Hamkins presents a similar thought experiment. He writes, "Imagine... that...the powerset size axiom [(PSA) that for any x and y, if |x| < |y|, then $2^x < 2^y$] had been considered at the very beginning of set theory...and was subsequently added to the standard list of axioms. In this case, perhaps we would now look upon models of ~PSA as strange in some fundamental way, violating a basic intuitive principle of sets concerning the relative sizes of power sets; perhaps our reaction to these models would be like the current reaction some mathematicians (not all) have to models of ZF+¬AC or to models of Aczel's

Strictly speaking, Pudlák is just pointing out what is uncontroversial—that the mathematical community could have had different set-theoretic beliefs, had history been different. But his point is evidently that that history *could have easily been different*. A world in which we rejected standard axioms is not that far off. Our belief that every non-empty set has a Choice function (AC) is simply not inevitable, even given the same evidence, standards of argument, level of intelligence, education, sincerity, attentiveness, and so on. As with moral disagreement, disagreement over AC often appears to bottom out in highly contingent "differences in…taste" ([Jensen 1995, 401]). "[T]he contestants," as Thomas Forster puts it, "have agreed to differ" ([Forthcoming, 58]). Even if those with a taste for AC happened to set the agenda for set theory—somewhat like those with a taste for internalism set the agenda for epistemology—we could have easily ended up skeptics of the orthodoxy. We could have easily studied with a different mentor, witnessed an enthralling talk by a heretic, or simply decided to go against the grain. Indeed, the orthodoxy *could still turn out to fail to be the orthodoxy*—in which case, as Hamkins puts it, the mathematical community generally may come to look at models of AC as "strange in some fundamental way."[44]

Of course, as I highlighted in Chapter 2, there are many differences between moral and mathematical disagreement. Our moral beliefs vary with our families and communities. By contrast, our mathematical beliefs are like academic beliefs more generally. They tend to depend on where we went to graduate school, who we studied with, and so forth. But it is hard to see how this could show that we could have easily had different moral, but not mathematical, beliefs. On the contrary, examples of paradigmatically academic beliefs—such as belief in epistemological internalism—are standard fare in the literature on "contingency anxiety" (Mogensen [2016]). Second, moral disagreement occurs among people of all levels of education and intelligence, while mathematical disagreement appears to be limited to the educated—again, like academic disagreement more generally. But, if anything, this makes it *easier* to argue that we could have easily had different mathematical beliefs in the requisite sense. It is much harder to deny the pertinent similarity of, say, Hugh Woodin and Ronald Jenson than it is to deny

anti-foundation axiom AFA, namely, the view that the models may be interesting mathematically and useful for a purpose, but ultimately they violate a basic principle of sets" [2011, 19].

[44] See Koellner, Bagaria, and Woodin [2017] for recent work on extremely large cardinals that are inconsistent with Choice.

the similarity of the multifarious parties to a typical moral disagreement. Finally, moral disagreement tends to track with personal and religious investment in a way that mathematical disagreement does not. However, again, this just makes it that much easier to explain it away as reflecting distorting factors—and to dismiss its participants as not appropriately alike.

5.9 Pluralism and Safety

So, prima facie, if *Moral Unreliability* is true, then *Mathematical Reliability* is false. Our mathematical beliefs also have a strong claim to being unsafe, realistically construed. So, if this interpretation of "explain the reliability" satisfies *Undermining Inexplicability*, it would appear that *neither* the moral *nor* the mathematical realist can explain the reliability of her beliefs in at least one sense which is such that the apparent impossibility of explaining their reliability may undermine them. But note that this conclusion depends on an assumption which has gone unmentioned. The assumption is that the mathematical realist is also an *objectivist* in the sense of Section 1.6—that is, that she does not believe that typical axioms of our foundational theories, like set theory, are analogous to the Parallel Postulate. (It also depends on the rarely questioned assumption that the moral realist is an objectivist in the sense of Section 1.6. I return to this assumption in Chapter 6.) If we jettison this, *and if we suppose that we could not have easily had inconsistent mathematical beliefs*, then *Mathematical Reliability* is plausible after all.[45] Without the objectivity assumption, the realist is free to be a *mathematical pluralist*—that is, to claim, at first pass, that every consistent mathematical theory is true of its intended subject, independent of human minds and languages. As I mentioned in Section 4.4, pluralism is widely taken to be the unique version of realism that affords an answer to the reliability challenge. Field himself says,

> [Some philosophers] (Balaguer (1995); Putnam (1980), perhaps Carnap (1950a)] (1983) solve the problem by articulating views on which though mathematical objects are mind-independent, any view we had had of them would have been correct.... [T]hese views allow for...knowledge in mathematics, and unlike more standard Platonist views, they seem to give an intelligible explanation of it. [2005, 78]

---

[45] This claim will require refinement in Section 6.2.

Field discusses such views in the context of (something like) sensitivity. But, as I mentioned in Section 4.3, pluralist views do nothing to show that our mathematical beliefs are sensitive (much less counterfactually persistent). Consider our belief that every set occurs at some level of the cumulative hierarchy. Then had this not been so, it seems that we still would have believed that it was (using the method that we actually used). Of course, assuming pluralism, the closest worlds in which some set fails to occur at some level of the cumulative hierarchy are worlds in which the "pluriverse" is different—just as, assuming objectivism, the closest worlds in which some set fails to so occur are worlds in which the universe is. For all that has been said, such worlds may be distant. But sensitivity does not care about distance. It just looks at the closest worlds in which the truth is different, however close they are, and checks what belief we form in them. If our belief that every set occurs at some level of the cumulative hierarchy is not sensitive, assuming objectivist realism, then there is no reason to think that it is sensitive assuming pluralism.

Nor does mathematical pluralism do anything to establish a causal, explanatory or even logical connection between our beliefs and the truths. *If mathematical pluralism answers any reliability challenge worth answering, it is the challenge to show that our mathematical beliefs are safe.* At first pass, she can argue thus. The mathematical truths could not have easily been different, because they are metaphysically necessary. Moreover, we could not have easily had *inconsistent* mathematical beliefs. So, by pluralism, we could not have easily had false ones.

The "so" will require more discussion. But the upshot is clear: if pluralism is tenable in the mathematical case, but not the moral, then there is, after all, a lack of parity between arguments for realism in the two areas. Only the mathematical realist can answer the reliability challenge.

5.10 Conclusions

I have discussed the reliability challenge for moral and mathematical realism. I have argued that it is the genuine challenge to which Genealogical Debunking Arguments point. I have substantially clarified Field's statement of it, and argued that it cannot be understood as the challenge to establish a causal, explanatory, or even counterfactual dependence between our beliefs and the truths, contrary to what both Benacerraf and Field suggest. Indeed, if it were understood as the challenge to show that our beliefs are sensitive, then it may well admit of an evolutionary answer in the moral case, but not in the mathematical—contrary to what is widely supposed. But this is just because we cannot establish the sensitivity of pretty much any of our beliefs in truths which fail to ascribe supervenient properties to concrete things, and can trivially establish the sensitivity of pretty much any of our beliefs that do ascribe those. The reliability challenge is better understood as the challenge to show that our beliefs are safe. Understanding the challenge in this way allows us to clarify the epistemological argument for mathematical pluralism, as well as the skeptical import of genealogy and disagreement. If the reliability challenge is understood in this way, however, then whether it is equally pressing in the moral and mathematical cases turns on whether realist pluralism is equally viable in the two areas.

It might be thought that I have overlooked a remaining way of understanding the reliability challenge, a formulation occasionally pressed by Field himself. In his [1989] work, he writes,

> If the intelligibility of talk of "varying the facts" is challenged... it can easily be dropped without much loss to the problem: there is still the problem of explaining the *actual* correlation between our believing "p" and its being the case that p.   [238, italics in original]

I do not know what this means. It might be taken to involve showing that the correlation holds in nearby worlds, so the actual correlation is no fluke. But, in that case, we are just back to something like safety or sensitivity.[46] Perhaps, then, there is a hyperintensional sense of "explanation" according to which one can intelligibly request an explanation of the "merely actual correlation" between our beliefs and the truths, and which is different from the sense invoked in *Answers 1–3*? Perhaps. But, even if there is, it would seem to apply equally in the mathematical case. Indeed, that was the case that Field was talking about. Moreover, as with the senses invoked in *Answers 1–3*, it is unclear how the apparent impossibility of offering such an explanation could *undermine* our beliefs. Even if we cannot explain the "merely actual correlation" between our moral or mathematical beliefs and the truths, in some hyperintensional sense of that phrase, we might still be able to show that those beliefs are sensitive, safe, and objectively probable, realistically construed (see, again, Section 4.5)! Finally, even if the above

---

[46] At the workshop on this book, Field humorously told me that he is not sure what it means either!

challenge can be made out, is distinct from any of those surveyed, and were worth taking seriously, as Field [2005, 78] suggests, pluralism may still afford the only (realist) answer to it.

Let me emphasize that I have not argued that the apparent impossibility of showing that our beliefs are safe *does* undermine them. I have remained neutral on this. What I have argued is that *if* there is an epistemological challenge which is such that (a) it appears impossible for the moral realist to answer, (b) it does not appear impossible for the mathematical realist to answer, and (c) the apparent impossibility of answering it undermines our moral beliefs, realistically construed, *then* it is the challenge to show that our beliefs are safe. It is possible that there simply is no sense of "explain the reliability" satisfying both (a) and (c).[47] Indeed, determining whether the apparent impossibility of showing that our beliefs of a kind are safe undermines them would seem to require determining how to individuate methods—and this is tantamount to resolving the notorious generality problem for process reliabilism (Conee and Feldman [1998]).

Suppose, then, for the sake of argument, that the apparent impossibility of showing that our beliefs of a kind, F, are safe does undermine them, realistically construed, and that the F-pluralist can show this. Is moral pluralism as viable as mathematical pluralism? In Chapter 6 I discuss this rarely considered question (but see Berry [2018] and Jonas [2019]), as well as the formulation of pluralism, in detail. The discussion finally reveals a deep difference between the cases. Unlike mathematical questions, moral questions, insofar as they are practical, do not answer to the facts.

[47] See my [2016] work for an argument that there is no intelligible reliability challenge. See Dogramaci [2016] for related discussion.

6

Realism, Objectivity, and Evaluation

I have argued that our mathematical beliefs have no better claim to being (defeasibly) a priori or a posteriori justified than our moral beliefs, realistically construed, that Genealogical Debunking Arguments are fallacious, and that whether the real problem they point to—the so-called reliability challenge—is equally pressing in the moral and mathematical cases depends on whether pluralism about morality and mathematics is equally viable. In this chapter I argue that even if moral pluralism per se is as viable as mathematical pluralism, pluralism about the practical questions at the center of our moral lives is not so much as intelligible. This shows that those questions are not moral, or even evaluative, questions, realistically construed. One upshot of the discussion is a radicalization of Moore's Open Question Argument. Another is that the concepts of realism and objectivity, which have been widely identified, are actually in tension.

6.1 Pluralism and Safety Again

Recall that mathematical pluralism is, *roughly,* the view that any consistent mathematical theory is true of its intended subject, independent of human minds and languages. I observed in Section 5.9 that pluralism is widely supposed to afford an answer to the reliability challenge for mathematical realism (that is, the challenge to explain the reliability of our mathematical beliefs). Indeed, this is its principal virtue. The following quotations are representative.

> [Pluralists] solve the problem by articulating views on which though mathematical objects are mind-independent, any view we had had of them would have been correct.... [T]hese views allow for ... knowledge in mathematics, and unlike more standard Platonist views, they seem to give an intelligible explanation of it.   [Field 2005, 78]

Morality and Mathematics. Justin Clarke-Doane, Oxford University Press (2020). © Justin Clarke-Doane.
DOI: 10.1093/oso/9780198823667.001.0001

The most important advantage that [pluralism] has over [objectivist] versions of platonism...is that all of the latter fall prey to Benacerraf's epistemological argument.   [Balaguer 1995, 317]

[The pluralist has] an answer to Benacerraf's worry that no link between our cognitive faculties and abstract objects accounts for our knowledge of the latter.   [Linksy and Zalta 1995, 25]

[The pluralist's] strong existence assumptions imply that, so long as we know that our full conception characterizing a theory is consistent, it can't fail to be true of...[a] portion of mathematical reality, so that knowledge of mathematical truths is reduced to knowledge of the consequences of consistent mathematical theories.   [Leng 2009, 124]

[Pluralism]...solve[s] the problem by expanding platonic heaven to such a degree that one's cognitive faculties can't miss it (as it were). (If you're having trouble hitting the target, then just make your target bigger!...).

[Beall 1999, 323]

If the reliability challenge for mathematical realism is understood as the challenge to show that our mathematical beliefs are safe, as I argued it should be in Section 5.8, then we can see why.[1] Roughly, the pluralist can argue as follows. First, with the objectivist realist, she can argue that the mathematical truths could not have easily been different, because they are metaphysically necessary. Second, while the objectivist must hold that we could not have easily had different, but still consistent, mathematical beliefs, the pluralist can allow that we could have. She can argue that, so long as our beliefs were consistent, they would have been true. Hence, so long as we could not have easily had inconsistent mathematical beliefs, we could not have easily had false ones (had we used the method that we actually used). Consequently, our mathematical beliefs are safe.

Why is this rough? Because the notions of "different" and "consistent" mathematical beliefs are ambiguous. Consider the mathematical *proposition*, P, such as that there is an inductive set. Then, whatever notion of consistency we take to be applicable to propositions, it had better be that ~P is

[1] Recall from Chapter 5 that it would be a mystery why mathematical pluralism is thought to be better placed to answer the reliability challenge if that challenge were any of the challenges with which it has traditionally been identified by itself advocates, like Benacerraf and Field—including the challenge to show that our beliefs are causally, explanatorily, or logically connected to the truths, or to show that they counterfactually depend on them.

consistent.[2] But no one can suggest that, given that P is true, had we believed ~P, our belief still would have been true. That is tantamount to P & ~P.[3] What the pluralist can hold is that had we accepted the *sentence* "it is not the case that there is an inductive set," which is *(first-order) consistent*, along with the sentence "there is an inductive set," that sentence would have expressed a different proposition out of our mouths, and that proposition is true too.[4]

Consider the Parallel Postulate *sentence*, S_{PP}.[5] It is not true that had we believed the negation of the *proposition* that we now use S_{PP} to express, our belief still would have been true. What is arguably true is that, had we uttered $\sim S_{PP}$, we would have expressed a different proposition, and that proposition is true right alongside the original (and remains true in nearby worlds).[6] If P_{PP} is the proposition expressed by S_{PP}, then, had we uttered $\sim S_{PP}$, we would have asserted a $\sim P_{PP}$-*like proposition*—where a $\sim P$-*like prop-osition* is the translation of ~P into a possibly distinct true proposition that intuitively shares ~P's "metaphysical content." The pluralist may claim that, rather than being a peculiar feature of geometrical postulates, this is the general situation. If the Axiom of Choice, understood as a proposition, is

---

[2] As it is according, for example, to Balaguer [1995].

[3] Assuming that ~P is true in this world if it is true in the closest world in which we believe it.

[4] I assume here that we have an independent grip on the notion of accepting a sentence. Balaguer [2009, 143] explicitly appeals to it, and Field [1998b, § 1] seems to require such a notion as well. Balaguer often uses a notion of consistency which applies to propositions, and even objects, in order to state his view. For instance, he writes, "[Pluralism] eliminates the mystery of how human beings could attain knowledge of mathematical objects. For if [pluralism] is correct, then all we have to do in order to attain such knowledge is conceptualize, or think about, or even 'dream up,' a mathematical object. Whatever we come up with, so long as it is consistent, we will have formed an accurate representation of some mathematical object, because, according to [pluralism], all possible mathematical objects exist" [1995, 317]. But I think that this is misleading. Does it seem inconsistent that the reals fail to be well-orderable (assuming that they are)? Certainly, "the reals fail to be well-orderable" has a model if ZF does, and Balaguer suggests [1995, § 3.5] that he accepts the extensional adequacy of the model-theoretic definition of consistency in the first-order case (although it is not clear what this means, since, again, Balaguer's notion of consistency applies to propositions, not sentences). But Balaguer holds that mathematical entities could not have existed while exemplifying different purely mathematical properties. So, if the *proposition* that the reals are well-orderable is true, then the proposition that the reals exist but fail to be well-orderable is not even consistent for Balaguer. This is understandable. Again, if it is consistent that the reals are well-orderable and consistent that they are not, then, according to Balaguer, the reals must be well-orderable and not! Perhaps the best way to think about Balaguer's semantics of consistency statements is on analogy with David Lewis's counterpart-theoretic semantics. Lewis grants that it is possible that Donald Trump could have been a ballerina. But he says that *what it is* for this to be so is, roughly, that a different person is one. My worry is then analogous to Saul Kripke's Humphrey objection to Lewis. See Kripke [1980, 45].

[5] See, again, Section 1.6 for a discussion of the Parallel Postulate.

[6] ~S denotes the negation of the sentence, S.

true "in" some part of the pluriverse, then it is not false in another. But the pluralist may hold that, along with sets, which are its subject matter, there are shmets. Shmets, we might say, are just like sets, except that some nonempty ones lack Choice functions—or "SChoice" functions. Again, the negation of the Parallel Postulate, *understood as a proposition about Euclidean space*, is simply false. But hyperbolic space is just like Euclidean space would be if the negation of the Parallel Postulate were true.

How, then, more carefully, can the pluralist answer the reliability challenge, *qua* the challenge to show that our mathematical beliefs are safe? Consider any mathematical proposition, P, that we believe. Then there are two ways that our belief that P could be unsafe. First, perhaps it could have easily been that ~P, while we still believed that P (using the method that we actually used to determine whether P). If mathematical realism is true, however, then this cannot be so when P is mathematical, because the mathematical truths could not have easily been different. Second, perhaps we could have easily believed that ~P—or, more generally, could have easily had a false belief as to whether Q, where Q is any proposition similar enough to P (using the method that we used).[7] But the pluralist holds that we could not have easily believed any such a Q either (using the method that we actually used to determine whether P) *so long as we could not have easily accepted (first-order) inconsistent mathematical sentences.* Her metasemantics is too cooperative for this. The consistent mathematical sentences that we accept are automatically about the parts of mathematical (or mathematical-like)[8] reality of which they are true, and there always are such parts. So, if mathematical pluralism is true, and we could not have easily accepted (first-order) inconsistent mathematical sentences, our belief that P is safe.[9]

6.2 Formulating Mathematical Pluralism

Let us define *pluralism* about an arbitrary area, F, as the view that there are a plurality of F-like concepts, all satisfied, independent of human minds and

---

[7] For the need to complicate the definition of safety in this way, see Section 4.6.

[8] If "set" actually refers to sets$_1$, say, then sets$_2$ are not technically sets—they are set-like things. I will not always add this qualification in what follows.

[9] I am not sure whether either Balaguer[1995] or Linsky and Zalta [1995] accept such a cooperative metasemantics, though, again, I think that it is hard to see the epistemological motivation for pluralism absent it. Field [1998b] does appear to accept the semantics. (Balaguer [2001] denies that our mathematical beliefs are exhausted by a simple—maybe even recursively enumerable—theory. But nothing I have said rules that out.)

languages.[10] More exactly, let F-pluralism be F-realism conjoined with the negation of F-Objectivity (Sections 1.2 and 1.6) and a cooperative metasemantics. Then pluralism comes in degrees, and not just any version of F-pluralism affords an answer to the reliability challenge for F-realism. In the mathematical case, one could imagine a view according to which there are exactly two set-like concepts that are satisfied—the cumulative hierarchical concept and Quine's of New Foundations (NF).[11] Then, while set-theoretic pluralism would officially be true, this would do little to show that our mathematical beliefs are safe. Assuming, for instance, that the Axiom of Choice is true "of" the cumulative hierarchical sets, I argued in Section 5.8 that we could have easily believed that it was false of them—or, at least, had a belief that was vacuously true of a third universe of set-like things, which is just like cumulative hierarchical sets except that some of them lack Choice functions (or "SChoice" functions). In order to answer the reliability challenge for realism, the pluralist must be radical.

Mathematical pluralism is widely formulated as radical indeed—as the view that *any (first-order) consistent* mathematical theory is true of its intended subject, independent of human minds and languages (Balaguer [1995], Field [1998a and 1998b], Hamkins [2012], Leng [2009]). This is *not* just the view that any consistent such theory has a model, which follows from the Completeness Theorem (Burgess [2001]]). The view is that any consistent such theory has an *intended* (class) model of the sort that ZF set theory is supposed by objectivist realists to have. But this formulation is untenable, at least if pluralism is supposed to show that it is *objectively* the case that we could not have easily had false mathematical (or, mathematical-like) beliefs.

The reason was broached in Sections 2.2 and 3.5. Consider any (first-order) consistent theory, T, interpreting Peano Arithmetic (PA). Then, if T is consistent, so is T conjoined with a coding of the claim that T is not consistent, T + ~Con(T), by Gödel's Second Incompleteness Theorem. So, if any consistent mathematical-like theory is true of its intended subject, then T + ~Con(T) is true of its. But that is tantamount to the view that the question of whether T is consistent is like the question of whether the Parallel

<hr>

[10] Note that a pluralist could technically hold that one set-like concept (or property) is "metaphysically privileged" (see, again, Section 1.6 as well as Section 6.5 below). But any reason to metaphysically privilege one concept of set over all others would seem to be a reason to regard only it as satisfied. Presumably that is why no one advocates this position, to my knowledge. (Whether one takes metaphysical privilege to attach to concepts or properties will be immaterial for my purposes. So, I will ignore the distinction between these views when the topic arises in the discussion that follows.)

[11] See, again, Boolos [1971], Shoenfield [1977], and Quine [1937], respectively.

Postulate is true. There is no objective fact as to whether there is a (classical) proof of a contradiction from the axioms of T. But, if this is our view, then even if we can argue that we could not have easily had false mathematical (mathematical-like) beliefs, we can argue equally for the intuitively opposite conclusion. Even if our belief in ZF is safe, because ZF is consistent, and we could not have easily had inconsistent set-theoretic beliefs, there is a symmetric argument that our belief in ZF is false, so certainly not safe, because ZF is inconsistent—or, rather, "shinconsistent." There is nothing objectively privileged about our notion of (classical) consistency. Related consequences of pluralism as standardly formulated include that there is no objective fact as to what counts as finite, as to what ZF *is*, and even as to what the language of ZF consists in. Surely this is too much to swallow![12]

A conservative way to moderate the above formulation of mathematical pluralism so as to avoid such consequences is to replace the (first-order) consistency requirement with a arithmetic soundness requirement. A theory is arithmetically sound when it does not imply a false sentence. If we replace the consistency requirement with a arithmetic soundness requirement, then only theories that are right about finiteness, consistency, and so on will count as true of their intended subjects. (Note that Con(PA), Con(ZF), and so forth are arithmetic, indeed Π_1 sentences.) This implies that the set of objective mathematical truths is no longer recursively enumerable, as it would be if the "pluriverse" witnessed every (first-order) consistent mathematical theory whatever. But it still appears to allow that any non-standard set (set-like) theory that we might have easily accepted is true of some part of the pluriverse. Presumably, we could not have easily believed the likes of ZF + ~Con(ZF), for example! For convenience, I will call a theory *coherent* when it is arithmetically sound.

Understood in this way, mathematical pluralism seems to be an intelligible position, and to have better claim to answering the reliability challenge, *qua* the challenge to show that our mathematical beliefs are safe, than any other version of realism. This is not incontrovertible, because it is not incontrovertible that we could not have easily had incoherent mathematical beliefs. But that we could not have is widely accepted, however implicitly. It is not obvious how to *argue* that we could not have. Field [1989, Introduction]

[12] See Chow [2017]. Thanks to Peter Koellner for convincing me years ago that the standard formulation of pluralism deserves scrutiny. But see Field [1994], Hamkins [2012], and Warren [2015] for attempts to make the view palatable. (Nothing turns on whether one agrees with me here. If one thinks that the pluralism as standardly formulated can be made sense of, then one is welcome to understand it that way in what follows.)

and Smith [2013, 13] suggest that if our mathematical beliefs were inconsistent, then a contradiction would have been discovered, and presumably
they would say something similar about an incoherence. But prima facie
the hypothesis that there is no proof of an incoherence from the theories in
question with fewer lines than there are atoms in the observable universe
explains our failure to derive an incoherence equally well (Leng [2007, § 4]).
Probably, any good argument that we could not have easily had incoherent
mathematical beliefs will have to be cumulative—noting not just mathematicians' failure to discover an incoherence, but scientists' enormous success
in applying mathematical theories to the concrete world, our insight into
models of those theories, and so on.

6.3 A Contrast

Mathematical pluralism is not an ad hoc response to the reliability challenge.
It has independent motivation. Recall from Sections 3.5 and 3.7 that there
seems to be no way to be a scientific realist without accepting some mathematics, realistically construed. So, if we wish to be scientific realists, we
must be mathematical realists. But mathematical objectivism generates
questions with no principled answer. Why should we privilege some one
notion of set, but no one notion of point or line? The pluralist avoids this
question. As Joel Hamkins writes,

> Today, geometers have a deep understanding of the alternative geometries,
> which are regarded as fully real and geometrical. The situation with set
> theory is the same. The initial concept of set put forth by Cantor and
> developed in the early days of set theory seemed to be about a unique
> concept of set, with set-theoretic arguments...seeming to take place in a
> unique background set-theoretic universe.... [But] today set theory is
> saturated with [alternative universes].   [2012, 426]

And Mark Balaguer notes,

> [I]f a mathematician comes up with a radically new [set] theory, she can
> be criticized on the grounds that the theory is inconsistent [incoherent]
> or uninteresting or useless, but she cannot be criticized on the grounds
> that the objects of the theory do not exist [or that the theory is otherwise
> false, but coherent]. Now, criticisms of this sort have emerged in the

history of mathematics (e.g., in connection with imaginary numbers) but, ultimately, they have never had any real effect...I think it is fair to say that...it is not a legitimate or interesting mathematical criticism to claim that the objects of a consistent purely mathematical theory do not exist [or that it is otherwise coherent, but false].   [1995, 311]

In general, mathematical pluralism says that all manner of intractable foundational questions—such as whether non-empty set "really" has a Choice function—are relevantly like the Parallel Postulate question.

What about moral pluralism? Metaphysically, moral pluralism is almost trivial. Recall from Section 1.5 that, unlike mathematics, morality at most "postulates" peculiar properties. It is *about*—i.e., names or (first-order) quantifies over—uncontroversial inhabitants of reality, like you and me. But properties come cheaply on a wide variety of conceptions, both Platonist and nominalist. This is in contrast to objects, including mathematical ones.[13] Virtually no one denies, for instance, that there is a property of maximizing utility (Bentham [1789, ch. 1], Mill [1863]), causally regulating our use of the word "good" (Boyd [1988]), or satisfying our "folk moral theory" (Jackson [1998]), in whatever sense there are properties at all.[14] (Of course, people deny that these properties are the actual referents of "good". But, then, people also deny that shmets are the actual referents of "sets". More on this below.) And while it might be thought that most philosophers at least deny that there are non-natural properties like those postulated by Moore [1903], Huemer [2005], or Enoch [2012], even this is far from evident. On typical formulations of Platonism about properties, properties are as abundant as can be. So, there are properties "up there" conforming to Moore's or Huemer's or Enoch's conceptions, right alongside Boyd's and Jackson's natural surrogates. On the other hand, the most influential formulation of nominalism about properties (Quine [1948]) says that the question of whether Moore's, Huemer's or Enoch's goodness exists is really just the question of whether to accept some primitive ideology into our regimented theory of the world. One could reasonably wonder how the predicate "is good" could be more objectionable than "is a restaurant."[15]

---

[13] But see Schiffer [2003] and Rayo [2013] for some exceptions to this rule.
[14] Nor does (virtually) anyone deny that these properties are instantiated.
[15] There is technically room for one very unusual kind of Aristotelianism about properties, according to which properties are sparse but include moral properties. The argument to follow that practical questions are not settled by the facts is independent of any particular conception of properties.

Nevertheless, there is clearly *something* unsatisfactory about moral, as opposed to mathematical, pluralism. At first pass: morality is supposed to tell us what to do. But moral pluralism leaves us clueless. While we can *believe* whatever theories we like (believe them true "of" different subjects), we can only do one thing. Knowing that we ought$_1$ to kill the one to save the five (in some situation), and ought$_2$ not leaves the *practical* question open: whether to.

6.4 The Irrelevance of Semantics

It is widely supposed that such considerations just show that the metasemantic component of moral pluralism is false. For instance, R. M. Hare uses the following thought experiment to challenge an analytic naturalism (he uses "relativism" instead of "pluralism"):

> [T]he naturalist kind of descriptivism leads inevitably to relativism.... [I]f we follow the naturalists, we shall have to say that the senses of the word in...two cultures are...different....If we distinguish the senses by using different subscripts, we can say that one of the cultures thinks fighting is wrong$_1$, but that the other thinks it is not wrong$_2$. But these two opinions may be mutually consistent, if the two senses of 'wrong' are different....   [1997, 30, italics in original]

Hare's point is that, according to analytic naturalism, cultures with very different moral-like views can both be right. They ascribe different properties— much like an advocate and a detractor of the Axiom of Choice, according to the mathematical pluralist. Horgan and Timmons [1992] make a similar point with their "moral twin earth" thought experiment. They point out that two cultures whose use of moral terms is causally regulated by different properties do not count as disagreeing according to "new wave" moral naturalism either. More generally, it might be thought that moral pluralism is false because it gets the metasemantics of moral language wrong—it predicts that people fail to have a moral disagreement when they really have one.

This would be a significant result. It would mean that mathematical realism is on a better epistemic footing than moral realism after all. The mathematical realist, but not the moral realist, can answer the reliability challenge, *qua* the challenge to show that our beliefs are safe. The mathematical realist, but not the moral realist, can be a pluralist (Berry [2018], Jonas [Forthcoming]).

But it cannot be that simple. Even if we all express the same property with "good," "ought," and so on, is it not enough that *there are* moral-like properties, good*, ought*, and so on giving intuitively opposite verdicts on the question of what to do? If we know that there are such things—who cares whether anyone's language latches onto them—then the question arises whether to regulate our behavior by consulting them, rather than the properties we actually consult.[16] Surely, the contingent fact of natural language semantics that we happen to latch onto ought$_1$ instead of ought$_2$ with "ought" should not settle the question of whether to kill the one![17]

Consider the set-theoretic case: even if we all refer to sets$_1$ with "sets," so that even undecidable sentences of ZF set theory, like Axiom of Choice (AC), have (determinate) truth-values, the mere existence of rival set-like universes giving intuitively opposite verdicts on such sentences would be enough to undercut the search for the true axioms. No mathematical pluralist would waste their time wondering whether AC is true, *assuming* the coherence of "every non-empty set has a Choice function" in tandem with the other axioms they accept. All they would learn in answering the question is something about them. They would just learn what parts of the set-theoretic pluriverse they were talking about, rather than learning what it contains (or what was "packed into" the concept of set they happened to be invoking). This is why (virtually) everyone agrees that *if* the metaphysical component of set-theoretic pluralism is true, *then* the question of what set-theoretic axioms are true is misconceived (Hamkins [2012])—while they also agree that, if the metaphysical component of set-theoretic objectivism is true, then that question is of paramount importance (Woodin [2001]).[18] Indeed, as I stressed in Section 1.6, the question of whether the search for new axioms to settle undecidables like the Continuum Hypothesis makes sense is just the question of whether such undecidables are relevantly like the Parallel Postulate. This is so *whatever the right metasemantics of set theory happens to be*. (Compare: even if we all happen to refer to the same

---

[16] I assume here that moral sentences ascribe properties, in accord with Moral-Face-Value from Chapter 1. But perfectly parallel points hold if one takes "ought" and so on to function differently, e.g., as operators.

[17] Eklund [Forthcoming] discusses the prospect of undercutting pluralism by appeal to a metasemantics according to which "normative role" determines reference (such as Wedgwood [2001]). But as he also appreciates, this just invites the follow-up question: what about the reference of normative-like roles?

[18] Assuming that the question is determinate, that is. See Section 5.2 and my [2013] work.

property with "simultaneous"—simultaneous-relative-to-reference-frame-R, say—this does nothing to show that there is a serious question as to whether two events were *really* simultaneous. And, yet, in this case, too, the question has a determinate mind-and-language independent answer out of our mouths.)

Such considerations should lead us to be suspicious that the problem with pluralism is just metasemantic, not metaphysical. If pluralism is problematic, when supplemented with a cooperative metasemantics, then it is problematic, qua metaphysical thesis, all on its own.

6.5 Radicalizing the Open Question Argument

I have argued that moral pluralism is unsatisfactory in a way that mathematical pluralism is not. At the same time, I have argued that it is hard to see how moral pluralism could be false—at least at a metaphysical level ,which is what seems to matter. However, there is a case to be made that the problem is worse. Arguably, it does not matter whether moral pluralism is true. It does not even matter whether it is metaphysically possible. Moral pluralism can be used to generate a problem for moral realism, *whether pluralist or objectivist*. The problem is that moral facts, even if there are any, would seem not to settle the practical question of what to do (Clarke-Doane [2015b]).[19]

One way of thinking about the "Open Question Argument" of Moore [1903, § 13] is that an agent may believe that A is F, for any *descriptive* property, F, while failing to endorse A in the sense that is characteristic of practical deliberation. She may grant that A is natural, or what she would desire herself to desire, or utility-maximizing, while still wondering whether to do it (and not merely in the sense that we all threaten to be weak in will). But why should it matter whether F is a "descriptive" property? As Simon Blackburn points out, "[e]ven if [a moral] belief were settled, there would still be issues of what importance to give it, what to do, and all the rest.... For any fact, there is a question of what to do about it" [1998, 70]. In other words, could not an agent believe that A is F, for any *moral* property, F, while failing to "endorse" A as well?

---

[19] See Eklund [Forthcoming] for a critical commentary on the argument as presented in my [2015b]. What follows significantly expands on that.

We can use moral pluralism to argue that one could (Clarke-Doane [2015b]). Let us *assume* that moral pluralism is true. We can either counter-factually conditionalize on it ("had it been the case that moral pluralism was true…"), or imagine that it "turns out" that moral pluralism is true, in the sense that it might turn out to be true that Hesperus ≠ Phosphorus. (It could surely turn out to be true! Again, by all appearances Boyd, Jackson, Scanlon, and others *actually are* pluralists. Perhaps we took a class in metaethics and came away convinced of Boyd's view, for instance.) Then while the assumption of mathematical pluralism *deflates* mathematical questions, the assumption of moral pluralism would not seem to deflate *practical* questions. The question of what to do would seem to remain open, even under the assumption of moral pluralism.

For example, the question of whether every non-empty set has a Choice function becomes analogous to the Parallel Postulate question under the assumption of pluralism. But the question of whether to kill the one to save the five is not deflated in this way. Granted that we ought$_1$ to kill the one, ought$_2$ not to, and so on, for any ought-like notions you like, the practical question of *whether to* kill the one remains open. The different ought-like notions "point" in different directions, leaving us with the practical question of which of them to follow. Even if one advocates bowing to the contingencies of natural language semantics—"following" the property that we happen to refer to with "ought"—this is a separate deliberative conclusion, not one that is out there in the moral pluriverse. An omniscient semanticist could not resolve the question of whether to kill the one *just* by confirming that we mean ought$_1$ by "ought"—so, indeed, we ought to kill the one (because we ought$_1$ kill the one)! It would appear that even Moore understated the case. Practical questions may remain open when the facts, *including the moral facts*, are settled.

We can make essentially the same point using the logical law of weakening (or, alternatively, the monotonicity of logical consequence). This says that if a conclusion, C, follows from premise A, then C certainly follows from premises A and B. Now, suppose that we know that, e.g., we ought to kill the one to save the five (in our present circumstance). Let us stipulatively introduce an ought-like concept, ought*, according to which we ought* not to kill the one to save the five. Then the question arises whether to do what we ought, or ought*, to do. Whatever *settling* a practical deliberation amounts to, it should at least satisfy the law of weakening in the following sense. If knowledge that we ought to kill the one *settles* the question of whether to kill the one on its own, then knowledge that we ought to kill the one settles this *in tandem with* knowledge that we ought* not to, by weakening.[20] Since it does not, knowledge that we ought to kill the one to save the five does not settle the question on its own. So, knowledge of the facts – even the moral facts -- fails to settle what to do.

Let me clarify two things about these arguments. First, their conclusion is not that *motivation externalism* is true. This says that we may sincerely judge that we ought to kill the one while failing to even be defeasibly motivated to do this (Brink [1986]). The conclusion is that our *deliberation as to whether* to kill the one is not completed even once we conclude that we *ought₁* to, that it is not the *thing to do₂*, that it would be *good₃* to, that we lack *reason₄* to, and so on, *for any moral-like properties you like*. We still need to ask whether to do what we ought₁ to do, is a thing to do₂, and so forth. Second, the conclusion is not just that knowledge of the moral facts, realistically construed, fails to settle what to do. It is that knowledge of the moral facts, *however construed*, fails to settle this -- so long as they are construed "cognitively" in the sense of Section 1.2. Consider, for example, a Korsgaardian constructivist who takes the question of whether we ought to kill the one to amount to that of whether this follows from our practical point of view. Although constructivists themselves emphasize the practical inertness of the realist's facts (Korsgaard [1996, 44], Street [2006, 138–9]), the facts that they posit are no less inert. For just as we can wonder whether to do what we ought, or ought*, to do, realistically construed, we can wonder whether to do what follows from our practical, or practical*, point of view. We can wonder, as Enoch [2006] puts it, whether to be an agent or a "shmagent."

6.6 Objections and Replies

There are various ways in which one might object to the above arguments— which I will often label the "New Open Question Arguments" on account of their affinity with Moore's. First, it might be responded that the New Open Question Arguments just shows that morality is not *overriding* (Das [Forthcoming]). They just show that sometimes we ought not *all-things-considered* do what we morally ought to do. Many moral realists would

---

[20] I should really say "weakening" because, as we will see, the conclusion is not a proposition, so the law does not strictly apply. Or, rather, if it applies, then "proposition" must be understood in a deflationary way (see Section 1.2 and 1.6). Note the appeal to *knowledge*, which is factive, not mere belief.)

happily agree. But, on the contrary, if the argument works, then it works for *any evaluative terms*—whether moral, prudential, epistemic, or aesthetic. For instance, even if we all-things-considered ought to kill the one to save the five, we all-things-considered ought* not (for some all-things-considered-like notion, ought*). And now the *practical* question arises whether to do what we all-things-considered ought, or all-things-considered ought*, to do. (This is why it does not help to maintain that this question just amounts to whether our moral-like "concepts adequately characterize the robustly normative properties" [Werner 2018, 9], whether they are "authoritatively normative" [McPherson [Forthcoming b], and so forth. If the New Open Question Arguments work, then they work equally for "normative," "authoritative," etc.)

Second, it might be objected that the argument just shows that we need to settle a question of *metaphysics* in order to settle our deliberation—namely, which of ought₁ or ought₂ is *metaphysically privileged* in something like the sense of Sider [2011] (McPherson and Enoch [2017]). Before I illustrate the problems with this objection, note that, if it were correct, then, again, moral realism would be on a worse epistemological footing than mathematical realism. Strictly speaking, the moral realist could answer the reliability challenge. She could argue that, since we could not have easily had incoherent moral-like beliefs, we could not have easily had false ones. But the moral pluralist at issue takes the epistemic good to be having true moral-like beliefs which ascribe metaphysically privileged properties, rather than merely taking it to be having true moral-like beliefs per se. So long as we could have easily had different moral-like beliefs, we could have easily failed to achieve the first good, even if we could not have easily failed to achieve the second. The apparent impossibility of answering *this* challenge would not seem to undermine our moral-like beliefs. Again, it is not that our moral-like beliefs would have been *false* in nearby worlds. But it would seem to undermine their practical clout. Even if we are right about the moral properties, maybe it is the moral* properties that are metaphysically privileged. The lesson would be that moral realism is on a worse footing than mathematical realism because the moral realist cannot answer something like the reliability challenge.

But this is not the lesson. Either the question of whether ought is metaphysically privileged is itself evaluative, or it is not. If it is not, then Moore's original Open Question Argument applies. Learning which of ought₁, ought₂, etc. is metaphysically privileged would be like learning which is brown. It would be neither here nor there from the standpoint of the practical question of which to use.[21] But if the question of whether ought is metaphysically privileged *is* evaluative, then, again, the present argument can just be rerun vis-à-vis privilege. Even if ought is not privileged, it is privileged*, for some alternative privileged-like concept, and the *practical* question arises whether to theorize with privileged or privileged* concepts (Dasgupta [2018]). (This is why it is not enough for the realist to say that "reality takes sides," even if it cannot carry us past the finish line.[22] Reality also takes sides*. And now the practical question is whether to care about sides or sides*.)

A third objection to the New Open Question Argument is that it trades on a false contrast. In the mathematical case, we noted that mathematical pluralism deflates mathematical questions, like whether every non-empty set has a Choice function. Such questions become like the Parallel Postulate question. But in the moral case, we said that moral pluralism fails to deflate *practical*—what to do—questions. We did not ask whether moral pluralism deflates moral questions per se (factually construed). And, indeed, it would seem to. This is arguably why debates in academic ethics, epistemology, and so forth threaten to be merely verbal, just like academic debates in non-evaluative fields such as speculative metaphysics. It is tempting to maintain, for instance, that the epistemic internalist and externalist can both be right—they are simply right about different properties (Alston [2005]). Given that mathematical pluralism also fails to deflate practical questions—such as which concept of set to use—there is no disanalogy after all.

But this objection is also misconceived. However similar mathematics and morality may be (and I have argued that their similarities are substantial),

---

[21] See Eklund [2017, 30], who uses the word "elite" instead of "metaphysically privileged," for a related point. He writes, "Suppose...that by eliteness one means something like what is sometimes called *Lewisian* eliteness: the perfectly elite properties are the fundamental physical properties, and something is more elite than something else the closer to this ideal it is. Then...[c]onsider two different communities with different aesthetic predicates—for example, they may have different, non-coextensive predicates 'tasty.' Suppose further that the tastes of one community are such that the extension of 'tasty' in their mouths is more metaphysically elite, for example because there is one particular chemical element, say *sodium*, such that they like—*gustatorily* like—food and drink that contains this chemical in sufficient quantities, so the referent of their 'tasty' is more elite. To take this to be relevant to which community objectively has the aesthetically better taste would clearly be unwarranted.... [T]he same goes for other normative disputes, including, for example, moral disputes, even if in other cases it is harder to come up with even prima facie compelling examples of greater eliteness in the Lewisian sense. The general take-home message is this: *even if what is more elite in the Lewisian sense may in some way be metaphysically privileged, it is not relevant so far as normativity or aesthetic evaluation is concerned*" [italics in original]. To be sure, one could just define a notion of evaluative privilege (or eliteness) such that our remaining question concerns which moral-like properties are privileged (van Roojen [2006, 180–1]). But that would make the proposal circular.

[22] Thanks to Matt Bedke for this way of putting it.

they are at least different in the following way. Mathematics is theoretical and morality is practical. We do not determine what we ought to do or believe for the sake of accumulating evaluative theorems. We do so to issue in action. But, then, the fact that knowledge of the moral and, more generally, evaluative facts fails to settle practical questions is quite damning. It does not show that moral realism is *false*. It shows that it fails to do the primary thing it should do—tell us what to do! By contrast, nobody would have suggested that pure mathematical facts would tell us—all by themselves—what to do, whether in the mathematical arena or elsewhere. They would not even tell us—by themselves—what to prove, or what axioms to use. That is just an instance of the Humean point that one cannot derive an "ought" from an "is," as well as an illustration of Moore's conclusion that one can know that something is F, for any *descriptive* property, F, while failing to "endorse" it.

Finally, one might argue that moral pluralism is simply unintelligible—we cannot conceive of moral-like properties. But it is hard to imagine a non-question-begging argument for this. Again, many prominent metaethicists – Boyd, Jackson, Scanlon, etc.—*actually are*, by all appearances, moral pluralists. It would certainly seem that their views are *intelligible*!

Let us suppose, however, that moral pluralism *is* unintelligible. *Why* would this be so? For typical descriptive areas, F, the notion of F-like properties makes sense. We can imagine set-like properties, grounding-like properties, possibility-like properties, essence-like properties, consequence-like properties, privilege-like properties, and so on. (Indeed, I will ultimately advocate pluralism about all of the corresponding areas.) If there are such things as moral properties, then why can we not imagine them "tweaked," just as we imagine the property of being a set tweaked? The obvious answer is that, in natural language, we do not use "ought to be done" to express a property at all. We use it to answer what-to-do questions.[23] *And pluralism about what to do does seem to be unintelligible.* But this truism is no thanks to special facts that we cannot even *assume* to be non-objective. It is thanks to the banal fact that we can only do *one thing*.[24]

Are there any other ways to object to the New Open Question Arguments? One could try to argue that the law of weakening is inappropriate in the

---

[23] This is exactly the moral that Blackburn draws from Moore's argument. He concludes, "evaluative discussion just is discussion of what to do about things" [1998, 70].

[24] Thanks to Jennifer McDonald for suggesting this way of putting the point. This response is especially compelling if Gibbard is right that the resolving attitude is intention (assuming that we cannot intend to X and ~X at the same time). More on this below.

present context, perhaps on the ground that it is invalid in some deontic logics. But even if this were right—and I think it is not—this would not get to the heart of the matter. For the original intuition, that practical questions are not deflated under the assumption of evaluative pluralism, would remain unexplained. Similarly, one could try to argue that the practical question of what to do is factual—it is simply ineffable.[25] But if this response is coherent, then we must be able to *mention* practical facts. And then the argument from pluralism can just be rerun at the level of practical pluralism.

There are two ways in which practical propositions could be ineffable. First, they could be structurally ineffable in the sense of Hofweber [2017]. Their ineffability could be due to their failure to share anything like sentential structure. But, if this were so, then it would be impossible to explain the connection between our linguistic behavior with moral sentences and the practical propositions we ponder. If you utter S and I reply ~S, where S is a moral sentence, then we should at least be able to conclude that the practical propositions that we believe are inconsistent (even if they are not expressed by the sentences, S, and ~S). But if practical propositions are structurally ineffable, then we do not even know whether "consistency" *makes sense* as applied to them—since we do not know whether there is any operation on them corresponding to sentential negation. So, it is more promising to suggest that practical propositions are ineffable because, while they share sentential structure, practical *properties* are ineffable. If this is why practical propositions are ineffable, however, then we could simply reformulate pluralism and bypass talk of sentences. Let us take *practical pluralism* to be the view, roughly, that all coherent practical-like propositions are true, albeit of different parts of the practical-like pluriverse. (We require structural effability in appealing to the notion of coherence here.) This would, in fact, be the analog of Balaguer's way of formulating mathematical pluralism. Then, even assuming practical pluralism, the question of whether to kill the one to save the five would seem to arise.

6.7 Realism, Objectivity, and Practical Safety

I have argued that knowledge of the facts—even the evaluative facts—fails to settle the practical questions at the center of our evaluative lives. Practical

[25] See Eklund [2017] for a response (which he does not endorse) to a related problem along these lines.

questions remain open even when all the facts, including the evaluative facts, are in. Of course, non-cognitivists like Gibbard [2003] could be right that, ordinarily, talk of evaluative "facts" is really just a way of expressing deliberative conclusions. Whether this is so is a question of natural language semantics. What is important is that evaluative facts *as the realist and, more generally, cognitivist, conceives of them* are practically anemic. This means that even if moral pluralism affords a resolution to the Benacerraf Problem for moral realism, understood as the problem of showing that our moral beliefs are safe (as mathematical pluralism seems to), there is a new problem of safety that moral pluralism fails to resolve (and which does not arise in the mathematical case). Where we might have worried that we could have easily had different, and so false, moral beliefs, we should now worry that we could have easily had moral*, rather than moral, beliefs. Had we, our moral-like beliefs may not have been *false* (they may have been true of the moral* facts). But we would have done what we would say in the vulgar we ought not to have (insofar as we were rational*). We would have been using the "wrong" evaluative, or evaluative-like, concepts (Eklund 2017). But this is *not* to say that we would have been using concepts which fail to be metaphysically privileged or carve at the joints. Again, metaphysical privilege, if it has ramifications for *good* theorizing, is itself an evaluative concept, and the argument from pluralism just reapplies to it (Dasgupta [2018]). The new problem of safety is not an *epistemological* problem at all. It is practical. But whereas the epistemological problem of safety was realists' alone, the new problem of safety is all of ours.

If the question of what to do is not settled by the facts, then what *would* settle it?[26] Settling it does not require *acting*, since we can resolve our deliberation as to whether to kill the one in the positive while failing to. But intending may not be enough either, contra Gibbard [2003]. Arguably, we can intend to do what we believe (to use the vulgar) we ought not to do.[27] That would make the elusive resolving attitude something in between acting and intending. Perhaps it can be informatively analyzed. Or maybe it must be taken as a primitive in moral psychology. But, again, the New Open Question Arguments would seem to show that we have it—whether or not moral, and more generally, evaluative realism is true. What matters is that

---

[26] Thanks to Geoffrey Sayre-McCord for pressing me on this.
[27] Thanks to Teemu Toppinen for helpful discussion of this point.

the attitude is not belief. The question of what to do remains after all of our beliefs are settled.[28]

On a traditional taxonomy, the conclusion of this chapter would be taken to show that practical questions are not objective. But we can now see how misleading an assessment this would be. *Realism* is false of those questions. But the concepts of realism and objectivity come apart. In fact, they are in tension.

Any mathematical realist concedes that there are independent geometrical facts, and that, whether a given geometrical sentence is true under the intended interpretation depends entirely on those facts. So, mathematical realists are geometrical realists as well. But pure geometry fails to be objective in a prototypical respect. If we found ourselves wondering whether the Parallel Postulate was true, understood as a pure mathematical conjecture, we could simply distinguish two concepts, lines$_{Euclidean}$ and lines$_{hyperbolic}$, and observe that it is true of the former and false of the latter. Or, at the intrapersonal level, a disagreement over the Parallel Postulate could be resolved by stipulation: you take lines$_{Euclidean}$ and I will take lines$_{hyperbolic}$. There is enough *mind-and-language independent* reality to go around. And while mathematical realists have traditionally contrasted areas like geometry with foundational areas like set theory, there is reason to be pluralists about those areas too. If there is one true set-theoretic universe, then we could have easily had false set-theoretic beliefs insofar as we could have easily had different ones. Pluralism, in tandem with a cooperative metasemantics, avoids this situation. It also avoids having to explain why we should privilege one set theory, while letting "a hundred flowers blossom" in geometry. The upshot is that if we are mathematical realists, then we should be pluralists. But this means that although Carnap was wrong to suggest that we can literally *generate* set-theoretic reality by specifying principles, he was right to remark that "the conflict between the divergent points of view...disappears...[B]efore us lies the boundless ocean of unlimited possibilities" [1937/2001, XV]. It is *as if* the most uncompromising relativism or constructivism were true.

In the other direction, an area may be objective in an important sense, though realism is false of it. Practical questions are highly objective in the sense in which the Parallel Postulate question is not. We cannot resolve them by saying "killing the one would be good$_1$ but bad$_2$, and that is all there is to it" or "you take good$_1$ and I will take good$_2$." In the practical

---

[28] See, however, Fitzpatrick [2008], Shafer-Landau [2009], and Sinclair [2007] for discussion of related phenomena from a realist point of view.

realm, we have to take a stand. Of course, many practical questions—such as those about which kind of shoes to wear—are not worth resolving. But they are still not vulnerable to deflation, in the way that the Parallel Postulate question is. Our conflict even over such trivialities may remain despite agreement on the facts, descriptive and evaluative. On the other hand, there are no practical facts. Practical questions are what remain when the facts, *even the evaluative ones*, come cheaply. Evaluative pluralism is almost trivial as a metaphysical thesis. But this does not deflate practical questions. In this sense, the objectivity of practical questions is robust. But it is robust *because* practical questions are not hostage to the facts. If they answered to the facts, then their objectivity would be compromised if the facts were abundant—just like set-theoretic questions. It follows that the concepts of realism and objectivity are not only independent. They are at odds.[29]

Note how elegantly this conclusion squares with the observation in Section 2.5 that mathematicians are overwhelmingly focused on the question of what is true if some axioms are, as opposed to the question of what axioms are true, while the situation seems roughly flipped in the moral case. The explanation, we can now see, is that morality, *insofar as it is practical*, is objective in a way that mathematics is not. In the mathematical case, we need not, in general, decide between apparently competing axioms. We need only agree on what is true if they are. By contrast, it is of almost no interest what follows from various moral principles per se. This is presumably why the idea of a "mathematical ethics" seems patently misconceived. Absent agreement over the "moral axioms," moral theorems are of negligible *practical* significance.

---

[29] One could try to make a similar argument to the "immunity to deflation" argument (but not the argument from weakening) for the conclusion that practical questions are not settled by the facts by conditionalizing on moral error theory, rather than moral pluralism. Perhaps this is even how we should read existentialists, like Sartre in his [1946] work. But this would require showing that we can believe error theory, which has been challenged. See Streumer [2013].

Conclusion

I have argued that the case for moral and mathematical realism is surprisingly parallel. Our mathematical beliefs have no better claim to being self-evident, provable, plausible, or analytic than our moral beliefs. Nor do our mathematical beliefs have better claim to being empirically justified than our moral beliefs. It is also incorrect that reflection on the genealogy of our moral beliefs undermines them, while reflection on the genealogy of our mathematical beliefs does not. What is true is that our moral beliefs are contingent in a worrying way, and this may disqualify them from counting as safe, realistically construed. However, exactly the same thing is true of our mathematical beliefs. So, contrary to the majority of the quotations that began this book, if one is a moral anti-realist on the basis of any such epistemological considerations, then one ought to be a mathematical anti-realist as well.

And, yet, moral realism and mathematical realism do not "stand or fall together," contrary to the following quotation from Hilary Putnam.

[A]rguments for "antirealism" in ethics are virtually identical with arguments for antirealism in the philosophy of mathematics; yet philosophers who resist those arguments in the latter case often capitulate in the former.   [2004, 1]

This is incorrect because a pluralist solution to the problem of safety is viable in the mathematical case, but not in the moral. Or, rather, if it is viable in the moral case, this is only because moral, and more generally evaluative, truths, fail to settle what to do. Practical questions—questions of what to do—are what remain after the facts, even the evaluative facts, come cheaply.

Gibbard asks:

When we answer fundamental questions of how to live, can our answers be objective in any important sense? Is the correctness of answers more than in the eye of beholders?   [2017, 748]

Morality and Mathematics. Justin Clarke-Doane, Oxford University Press (2020). © Justin Clarke-Doane.
DOI: 10.1093/oso/9780198823667.001.0001

I have argued that these questions come apart. The correctness of answers to mathematical questions is not just in the eye of the beholder—it depends on how the mind-and-language independent mathematical facts are. But those answers fail to be objective in a paradigmatic sense. In a disagreement over the Axiom of Choice, we can both be right. We can simply be right of different subjects. On the other hand, if we found ourselves disagreeing over practical questions, then our disagreement would not admit of such an ecumenical resolution. In a disagreement over whether to kill the one to save the five, we must either kill the one or not. But the correctness of practical judgments may only be in the eye of the beholder. In a slogan: while realism (in the sense of Section 1.2) is true of mathematics, it is not objective (in the sense of Section 1.6). And while morality, insofar as it is practical, is objective, realism is false of it.

C.1 Key Themes

Before turning to the broader relevance of this conclusion, let me review some of the upshots of the book. The first is that there is no *epistemological* ground on which to be a moral anti-realist and a mathematical realist. Even if the New Open Question Arguments (Section 6.4) succeed against moral—or practical—realism, all of the epistemological arguments against moral realism work equally against mathematical realism, whether from disagreement, dispensability, or lack of safety. So, absent another argument for moral anti-realism which lacks an analog in the mathematical case, one must be a mathematical anti-realist if one is a moral anti-realist *if one rejects the New Open Question Arguments* — contrary to what is widely assumed.

A second upshot is that the extent of disagreement over a topic may be of little epistemological consequence. We tend to work with cartoons of the established sciences, where agreement prevails, and of philosophy, where people argue endlessly. At the top of the pyramid is pure mathematics, where outright consensus is the norm, since mathematical propositions admit of proof. But we have seen that this picture is misleading. First, there is nothing special about mathematical proof. It is open to all of us, whether ethicists, chemists, or astrologers. Mathematical proof at most establishes that one claim follows from another—and it does not even establish that in a context in which the logic used is in doubt (as is the case when, say, an intuitionist is in the room). Second, among those vanishingly few who have a serious position on the question of what mathematical axioms are true, it is false that there is consensus, and if anyone has an algorithm for settling the disputes, I am not aware of it. Third, the case of mathematics may be illustrative of the general case. At the bottom of every established science are the strict and literal "philosophical" questions which are debated endlessly. To pretend that there is agreement over mathematics in the absence of agreement over philosophy of mathematics is to pretend that mathematics *in the epistemologically important sense* is not theoretical. While two mathematicians might agree to *use* the Axiom of Choice, this does nothing to show that they agree on that axiom in the sense that should matter to epistemologists. Contrary to the cartoon of established science, all areas of inquiry may be contentious, in the important sense, because all may bottom out in controversial philosophy.

A third upshot is additional reason for skepticism about a useful a priori/a posteriori distinction. Quine [1951b, § VI], of course, criticized the distinction on the grounds that no claim is immune to revision. But even assuming that a priori justification is defeasible, there appears to be no principled way to characterize experience from which a priori justification is supposed to be independent. It is frequently assumed that one can partition the a priori and the a posteriori in terms of subject matter—the former concerns abstract objects, while the latter concerns concrete ones (Bengson [2015], Bonjour [1997], Chudnoff [2013, Introduction], Lewis [1986, 108–15]). But morality, interpreted at face value, does not concern abstract objects. It concerns the likes of you and me. Neither does modality if modal operators are taken as primitive. To assume otherwise is to assume realism about universals, and risk making *all* of our beliefs qualify as a priori. And while one might appeal to "conceptual competence" to distinguish the spontaneous judgments about concrete things that count as based on experience, the view that *non-trivial* moral, mathematical, or modal belief stems from conceptual competence is an article of faith.

A fourth upshot is that the epistemological relevance of indispensability considerations has been widely misunderstood by metaethicists and philosophers of mathematics alike. It has been assumed that whether we can explain the reliability of our belief that P turns on whether P is implied by some explanation of our coming to believe that P. Apparently, philosophers have noticed that when P corresponds to a *causally efficacious* fact, such an explanatory connection is predictive of epistemically valuable features, like sensitivity, safety, or objective probability. However, no such prediction is justified when P would be causally inert, as mathematical and moral facts are supposed to be. In order to explain the reliability of our mathematical beliefs, it does not suffice to show that their contents are implied by some explanation of our coming to have them, and in order to debunk our moral beliefs it does not suffice to show the opposite.

A fifth upshot is that the sensitivity condition has limited application to the reliability challenge, and to Genealogical Debunking Arguments, properly formulated. The problem is *not* that counterfactuals conditionalizing on metaphysically impossible antecedents are vacuous. (And the problem with the view that they *are* is not that a non-standard semantics for counterfactuals is true. It is that metaphysical possibility is not maximal. See Section 3.4.). The problem is that the condition is trivial to meet whenever the beliefs in question predicate supervenient properties of concrete things, and impossible to meet whenever their contents are metaphysically necessary. This implies, for instance, that we can explain the reliability of our atomic astrological beliefs—and, indeed, may even be able to show that we were selected to have true ones per se!—but cannot explain the reliability of our mathematical, or perhaps even metalogical, beliefs. Our focus should be on the safety condition. Arguments from evolution and disagreement can both be understood as special cases of arguments from lack of safety.

A sixth upshot is a partial vindication of the Kroneckerian view that "God created the integers; all else is the work of man." More exactly, God seems to have created a *unique* set of (positive) integers, but myriad variations on other structures. We saw that a fragment of objective arithmetic is not only indispensable to science but even for intelligibly framing the reliability challenge. One can verbally disagree with this assessment by shunning numbers in favor of a primitive consistency operator. But nothing is gained by this approach. It just trades ontology for ideology, and whatever epistemological problems plague knowledge of objects will follow the trade. On the other hand, so long as we could not have easily had arithmetically unsound mathematical beliefs, the amount of arithmetic objectivity required arguably fails to threaten the safety of our mathematical beliefs.

A seventh upshot is that standard methodology in normative philosophy deserves renewed scrutiny. A common practice (illustrated in, for example, Cuneo and Shafer-Landau [2014]) is to arrive at moral, epistemic, prudential, and political conclusions by appeal to our evaluative concepts. We consider counterfactual cases, and ask what we would say. But if the arguments here are compelling, then relying on the concepts we happen to have inherited is no less conservative than relying on the evaluative beliefs we happen to have inherited.[1] Even if conceptual surgery reveals that our concept of responsibility entails retribution, for example, this does not settle the practical question of whether to retribute. That depends on whether to hold people responsible rather than responsible*, for some responsibility-like concept, responsibility*, according to which responsibility* does not entail retribution (or retribution*). For any evaluative concepts, we can "critique…the value of these values" [Nietzsche 1887, Preface, § 6].

A final upshot is that we need to fully distinguish two concepts that have been widely identified—realism and objectivity. We have seen that one can be a realist about a domain, F, while rejecting the objectivity of the F-truths. This is the position of the mathematical pluralist, and it is the position of virtually all of us about (pure) geometry. But one can also be an F-objectivist without being an F-realist. This is the position of the anti-realist who denies that practical—what-to-do—questions are hostage to the facts, as well as the position of an if-thenist about mathematics who takes modal operators as primitive (a position that I will illustrate presently). Which notion should be philosophers' focus? Let me turn to that question now.

C.2 Toward Practical Philosophy

It is sometimes suggested that the question of objectivity, rather than objects, is what matters methodologically (Field [1998a]). If mathematics is not objective, then the search for new axioms to settle undecidables like the Continuum Hypothesis, as traditionally conceived, is misguided—*even if* mathematical realism is true. On the other hand, if a mathematical sentence "S" were shorthand for the claim that it is mathematically necessary that S, then that search may *not* be misguided, even though mathematical realism would be false. It would not be misguided if it was mathematically necessary that S just when "S" was true in the one true universe of sets, V, according to the objectivist realist. Similarly, if paradigmatic questions of philosophy, such as the question of whether we are free, whether a brain in a vat is justified, or whether the grounding relation is transitive, turn out not to be objective, then the search for their answers, as traditionally conceived, is misguided as well. This is so *even if philosophical realism is true.*

---

[1] Eklund also discusses the conservativeness of relying on our actual normative concepts in his [2017] work. See Dutilh Novaes [2015] for related discussion.

Consider the question of whether it is possible that you had different parents. Kripke [1980, 113] famously argued that it is not. But now suppose that what I said in Section 3.4 about metaphysical possibility is correct. It is not maximal possibility, even by Timothy Williamson's lights. Moreover, there is nothing to bar us from defining all manner of more and less inclusive notions of possibility, even holding fixed the modal logic. So long as we read "<>" to mean that *it could have been the case that*, we can consider scenarios where water (that very thing!) fails to be composed of H_2O, where it is morally obligatory to count blades of grass, and so on. Indeed, scenarios such as these are already studied under the heading "impossible worlds" ([Berto 2013]). We can also, of course, limit ourselves to scenarios that respect the metaphysical or even physical laws. So, let us ask again: could we have had different parents? The question, *even if it admits of a determinate mind-and-language independent answer out of our mouths*, is not objective. Relative to some notions of possibility, we could have, and relative to those weaker than so-called metaphysical possibility, we could not have. There is no deeper answer.

It might be thought that the objectivity of modal questions could be vindicated by appeal to another notion—the notion of nature. Metaphysical necessities, unlike necessities in this stricter sense, are those that are grounded in the natures of things (Hale [2013], Lowe [2012], Fine [1994], Kment [2014]). But a skeptic about the objectivity of modality should just be a skeptic about the objectivity of nature. In addition to nature, let us consider nature*. Just as people have natures, they have natures*. But while it is part of the nature of people that they have the parents that they have, this is no part of their nature*. Absent a direct argument for the objectivity of modal questions, any argument for the objectivity of nature questions would seem to require appealing to yet another notion whose objectivity should be just as much in dispute.

More credibly, I suggest, different notions of possibility, and different notions of nature, are like different notions of geometrical point or line—or, in the context of set-theoretic pluralism, like different notions of set. While we take interest in some over others, the question of what is possible, or what is part of a thing's nature, is no more objective than the Parallel Postulate question. For practical purposes, the question of whether we could have had different parents is "merely verbal."[2]

[2] See Cameron [2009] and Sider [2011, ch. 12] for alternative deflationary conceptions of modal questions.

Analogous reasoning suggests a more general metaphilosophical deflationism about apparently a priori descriptive questions. Consider logic understood as a factual area. According to classical logic, it follows from the premise that snow is both white and not white that we are all fish. According to paraconsistent logics, it does not. Does it *really* follow? Of course, we speak a language, and for all that has been said, we may determinately mean classical consequence by "follows from." In that case, the question may have a determinate mind-and-language independent answer out of our mouths. But the question of whether classical logic is objectively correct is not usefully characterized as that of whether "the implication relation so defined agrees with the pre-theoretic notion of implication between statements" [Zach Forthcoming, 1]. By that standard, Euclidean geometry may be objectively correct! Now contrast the factual question of what follows from what with the practical question of *whether to* believe that, for example, we are all fish if we fully believe that snow is both white and not white. That question, however bizarre, cannot be deflated in the same way. It is a practical question, and such questions are impervious to how plentiful the facts turn out to be.

Practical questions, whether epistemic, prudential, or moral, are not going away. I do not deny what Parfit [1984, 454] suggests, that we might reach consensus on some of them. My point is that this would be a practical achievement, not a terminological victory of the sort we would secure if everyone were to agree that the Parallel Postulate is true, understood as a pure mathematical conjecture. My own view is that pluralism about grounding, nature, mereology, logic, metaphysical privilege, and much more are the most viable forms of realism about the corresponding areas. If so, then questions characteristic of them are all like the Parallel Postulate question. In calling them "verbal," critics would be wrong in the letter but right in the spirit.[3]

But that does not mean that there are no non-verbal questions in the neighborhood of the above questions. On the contrary, there are practical questions. Even a set-theoretic pluralist, who concedes that there is no serious question as to whether the Continuum Hypothesis (CH) is true, can, of

---

[3] Let me mention a complication for the picture I am sketching here which I hope to address in future work. Pluralism about metaphysical privilege leads to pluralism about likeness (in the sense of Section 1.6), because it leads to pluralism about natural kinds. And this raises the specter of pluralism about pluralism. Indeed, if our belief in, say, mathematical pluralism would itself be unsafe if non-pluralist realism about pluralism (!) were true, then I would seem to be committed to pluralism about pluralism, independent of my pluralism about privilege. This is a version of the self-reference problem for philosophical deflationism.

course, inquire into what concept of set to use. Indeed, "debates" about new axioms are often explicitly practical in this way (Bagaria [2005]). To frame the question as one about the extension of "is a member of," or about what is packed into our concept of set, or even about what the set-theoretic universe contains, just obscures the question that is really at stake.[4]

Of course, if mathematical pluralism is *false*, then there is a serious question of whether CH is true. There is, in this case, one true mathematical world, just as there is one true physical world, and we should know it. The problem is that, in that case, it is doubtful that we could. The reliability challenge would be as pressing as ever. Although pluralism is methodologically similar to Carnap [1950a], it does not pretend to step outside of metaphysics. Unlike Carnap, I do not believe that this is possible. Pluralism, whether mathematical, mereological, or logical, is just more metaphysics, and, as such, is itself vulnerable to all the familiar skeptical doubts.

---

[4] It is striking that critics of philosophy almost always make an exception for practical subjects, like normative ethics. For instance, Peter Unger in his book-length critique of analytic philosophy begins, "my argumentation *won't* concern anything that's deeply normative, or fully evaluative, or anything of the ilk. Most certainly, I won't seek to address any claims that are obviously, or explicitly, or paradigmatically of any such sorts or kinds" [2014, 4]. If the arguments here are compelling, then we can see why. Normative questions, insofar as they are practical, are objective in a way that other paradigmatically philosophical questions are not.

Bibliography

Alston, William. [2005] *Beyond "Justification": Dimensions of Epistemic Evaluation*, Ithaca, NY: Cornell University Press.

Arntzenius, Frank, and Cian Dorr. [2012] "Calculus as Geometry," in Arntzenius, Frank (ed.), Space, Time and Stuff. Oxford: Oxford University Press.

Arrigoni, Tatiana. [2011] "V=L and Intuitive Plausibility in Set Theory. A Case Study." *Bulletin of Symbolic Logic*. Vol. 17. 337–59.

Ayer, A. J. [1936] *Language, Truth, and Logic*. London: Victor Gollancz Ltd.

Azcel, Peter. [1988] *Non-Well-Founded Sets. SLI Lecture Notes. Vol. 14*. Stanford, CA: Stanford University, Center for the Study of Language and Information. Available online at http://www.irafs.org/courses/materials/aczel_set_theory.pdf.

Bagaria, Joan. [2005] "Natural Axioms of Set Theory and the Continuum Problem," in Hajek, Petr, Luis Valdes-Villanueva, and Dag Westerstahl (eds.), Logic, Methodology and Philosophy of Science. Proceedings of the Twelfth International Congress. London: King's College Publications. 43–64.

Baker, Alan. [2003] "Does the Existence of Mathematical Objects Make a Difference?" *Australasian Journal of Philosophy*. Vol. 81. 246–64.

Baker, Alan. [2005] "Are there Genuine Mathematical Explanations of Physical Phenomena?" *Mind*. Vol. 114. 223–38.

Balaguer, Mark. [1995] "A Platonist Epistemology." *Synthese*. Vol. 103. 303–25.

Balaguer, Mark. [1998] *Platonism and Anti-Platonism in Mathematics*. New York: Oxford University Press.

Balaguer, Mark. [1999] "Review of Michael Resnik's Mathematics as a Science of Patterns." *Philosophia Mathematica*. Vol. 7. 108–26.

Balaguer, Mark. [2001] "A Theory of Mathematical Correctness and Mathematical Truth." *Pacific Philosophical Quarterly*. Vol. 82. 87–114.

Balaguer, Mark. [2009] "Fictionalism, Theft, and the Story of Mathematics." *Philosophia Mathematica*. Vol. 17. 131–62.

Balaguer, Mark. [2016] "Platonism in Metaphysics," in The Stanford Encyclopedia of Philosophy (Spring Edn), Edward N. Zalta (ed.), https://plato.stanford.edu/archives/spr2016/entries/platonism/.

Bangu, Sorin. [2008] "Inference to the Best Explanation and Mathematical Realism." *Synthese*. Vol. 160. 13–20.

Baras, Dan. [2017] "Our Reliability is in Principle Explainable." *Episteme*. Vol. 14. 197–211.

Baras, Dan, and Justin Clarke-Doane. [Forthcoming] "Modal Security." *Philosophy and Phenomenological Research*.

Barton, Neil. [2016] "Multiversism and Concepts of Set: How Much Relativism is Acceptable?" in Francesca Boccuni and Andrea Sereni (eds.), *Objectivity, Realism, and Proof*. Springer. 189–209.

Bealer, George. [1982] *Quality and Concept*. Oxford: Clarendon Press.

Bealer, George. [1999] "A Theory of the A Priori." *Philosophical Perspectives*. Vol. 13. 29–55.

Beall, J. C. [1999] "From Full-Blooded Platonism to Really Full-Blooded Platonism." *Philosophia Mathematica*. Vol. 7. 322–7.

Bedke, Matthew. [2009] "Intuitive Non-Naturalism Meets Cosmic Coincidence. *Pacific Philosophical Quarterly*. Vol. 90. 188–209.

Bell, John, and Geoffrey Hellman. [2006] "Pluralism and the Foundations of Mathematics," in Waters, Kenneth, Helen Longino, and Stephen Kellert (eds.), *Scientific Pluralism (Minnesota Studies in Philosophy of Science, Volume 19)*. Minneapolis: University of Minnesota Press.

Benacerraf, Paul. [1965] "What Numbers Could Not Be." *Philosophical Review*. Vol. 74. 47–73.

Benacerraf, Paul. [1973] "Mathematical Truth." *Journal of Philosophy*. Vol. 70. 661–79.

Bengson, John. [2015] "Grasping the Third Realm," in Szabó Gendler, Tamar, and John Hawthorne (eds.). *Oxford Studies in Epistemology, Vol. 5*. Oxford: Oxford University Press.

Bentham, Jeremy. [1789] *An Introduction to the Principles of Morals and Legislation*. London: T. Payne & Son. Available online at http://www.koeblergerhard.de/Fontes/BenthamJeremyMoralsandLegislation1789.pdf.

Berry, Sharon. [2018] "(Probably) Not Companions in Guilt." *Philosophical Studies*. Vol. 175. 2285-2308.

Berry, Sharon. [Manuscript] "Coincidence Avoidance and Formulating the Access Problem." Available online at http://seberry.org/coincidence-avoidance.pdf.

Berto, Francesco [2013] "Impossible Worlds," in The Stanford Encyclopedia of Philosophy (Winter Edn), Edward N. Zalta (ed.), https://plato.stanford.edu/archives/win2013/entries/impossible-worlds/.

Blackburn, Simon. [1971] "Moral Realism," in Casey, John (ed.), *Morality and Moral Reasoning*. London: Methuen.

Blackburn, Simon. [1984] *Spreading the Word*. Oxford: Clarendon Press.

Blackburn, Simon. [1993] Essays in Quasi-Realism. Oxford: Oxford University Press.

Blackburn, Simon. [2008] *The Oxford Dictionary of Philosophy*. Oxford: Oxford University Press.

Boghossian, Paul. [1997] "Analyticity," in Hale, Bob, and Crispin Wright (eds.), *A Companion to the Philosophy of Language*. Oxford: Blackwell. 331–68.

Boghossian, Paul. [2003] "Epistemic Analyticity: A Defense." *Grazer Philosophische Studien*. Vol. 66. 15–35.

Bonjour, Laurence. [1997] *In Defense of Pure Reason: A Rationalist Account of A Priori Justification*. Cambridge: Cambridge University Press.

Boolos, George. [1971] "The Iterative Conception of Set." *Journal of Philosophy*. Vol. 68. 215–31.

Boolos, George. [1999] "Must We Believe in Set Theory?" in *Logic, Logic, and Logic*. Cambridge: Harvard University Press.

Boyd, Richard. [1988] "How to Be a Moral Realist," in Geoffrey Sayre-McCord (ed.), *Essays on Moral Realism*. Ithaca: Cornell University Press. 181–228.

Braddock, Matthew, Walter Sinnott-Armstrong, and Andreas Mogensen. [2012] "Comments on Justin Clarke-Doane's 'Morality and Mathematics: The Evolutionary Challenge." *Ethics at PEA Soup*. Available online at http://peasoup.typepad.com/peasoup/2012/03/ethics-discussions-at-pea-soupjustin-clarke-doanesmorality-and-mathematics-the-evolutionary-challe-1.html.

Brink, David. [1986] "Externalist Moral Realism." *Southern Journal of Philosophy*. Vol. 24. (Supplement): 23–40.

Brink, David. [1989] *Moral Realism and the Foundations of Ethics*. Cambridge: Cambridge University Press.

Brouwer, L. E. J. [1983/1949] "Consciousness, Philosophy and Mathematics," in Benacerraf, Paul, and Hilary Putnam (eds.), *Philosophy of Mathematics: Selected Readings (2nd edn)*.Cambridge: Cambridge University Press.

Brown, Campbell [2011] "A New and Improved Supervenience Argument for Ethical Descriptivism," in Russ Shafer-Landau (ed.), *Oxford Studies in Metaethics, Vol. 6*. Oxford: Oxford University Press. 205–18.

Brown, James Robert. [2019] "Ethics and the Continuum Hypothesis," in Fillion, N., Corless, R. M. C., Kotsireas, I. (eds.), *Algorithms and Complexity in Mathematics, Epistemology, and Science*. New York: Springer. 1–16.

Burgess, John. [2001] "Review of Platonism and Anti-Platonism in Mathematics." *Philosophical Review*. Vol. 110. 79—82.

Burgess, John. [2005] Fixing Frege. Princeton: Princeton University Press.

Burgess, John, [2007] "Against Ethics." *Ethical Theory and Moral Practice*. Vol. 10. 427–39.

Burgess, Alexis, and David Plunkett. [2013a] "Conceptual Ethics I." *Philosophy Compass*. Vol. 8. 1091–101.

Burgess, Alexis, and David Plunkett. [2013b]"Conceptual Ethics II." *Philosophy Compass*. Vol. 8. 1102–10.

Burgess, John, and Gideon Rosen. [1997] *A Subject with No Object: Strategies for a Nominalistic Interpretation of Mathematics*. Oxford: Clarendon Press.

Burnyeat, Myles. [2000] "Plato on Why Mathematics is Good for the Soul," in Smiley, Timothy (ed.), *Mathematics and Necessity: Essays in the History of Philosophy*. Oxford: Oxford University Press. 1–81.

Butterworth, Brian. [1999] *What Counts? How Every Brain is Hardwired for Math*. New York: The Free Press.

Cameron, Ross. [2009] "What's Metaphysical about Metaphysical Necessity?" *Philosophy and Phenomenological Research*. Vol. 79. 1–16.

Carnap, Rudolf. [1950a] "Empiricism, Semantics, and Ontology." *Revue Internationale de Philosophie*. Vol. 4. 20–40.

Carnap, Rudolf. [1950b] *The Logical Foundations of Probability*. Chicago: University of Chicago Press.

Carnap, Rudolf. [1999] "Abraham Kaplan on Value Judgments," in Schillp, Paul (ed.). *The Philosophy of Rudolf Carnap, Vol. 11 (Library of Living Philosophers)*. La Salle, IL: Open Court.

Carnap, Rudolf. [2001/1937] *The Logical Syntax of Language*. Oxford: Routledge.

Carroll, Sean. [2010a] "The Moral Equivalent of the Parallel Postulate." Post on *Cosmic Variance* blog (Discover magazine), March 24, 2010. Available online at http://blogs.discovermagazine.com/cosmicvariance/2010/03/24/the-moral-equivalent-of-the-parallel-postulate/#.Wz7F39JKg2w.

Carroll, Sean. [2010b] *From Eternity to Here: The Question for the Ultimate Theory of Time*. London: Penguin.

Casullo, Albert. [2002] "A Priori Knowledge," in Moser, Paul (ed.), *Oxford Handbook of Epistemology*. Oxford: Oxford University Press. 95–143.

Casullo, Albert. [2011] "Reply to My Critics: Anthony Brueckner and Robin Jeshion," in Shaffer, Michael, and Michael Veber (eds.), *What Place for the A Priori?* La Salle, IL: Open Court Press.

Chang, Ruth. [2004] "All Things Considered," in Hawthorne, John, and Dean Zimmerman (eds.), *Ethics, Philosophical Perspectives*. Vol. 18. Oxford: Wiley-Blackwell. 1–22.

Chen, Eddy Keming. [2018] "The Intrinsic Structure of Quantum Mechanics." *PhilSci Archive*. Available online at http://philsci-archive.pitt.edu/15140/.

Cheyne, Colin. [1998] "Existence Claims and Causality." *Australasian Journal of Philosophy*. Vol. 76. 34–47.

Chihara, Charles. [1990] *Constructability and Mathematical Existence*. Oxford: Oxford University Press.

Chow, Timothy. [2017] "Re: [FOM] Hamkins's Multiverse and Ultrafinitism." Post on *Foundations of Mathematics (FOM)*. November 29, 2017. Available online at https://cs.nyu.edu/pipermail/fom/2017-November/020701.html.

Chudnoff, Elijah. [2013] *Intuition*. Oxford: Oxford University Press.

Clarke, Samuel. [2010/1705] "A Discourse of Natural Religion," in Nadelhoffer, Thomas, Eddy Nahmias, and Shaun Nichols (eds.), *Moral Psychology: Historical and Contemporary Readings*. Oxford: Wiley-Blackwell.

Clarke-Doane, Justin. [2008] "Multiple Reductions Revisited." *Philosophia Mathematica*. Vol. 16. 244–55.

Clarke-Doane, Justin. [2012] "Morality and Mathematics: The Evolutionary Challenge." *Ethics*. Vol. 122. 313–40.

Clarke-Doane, Justin. [2014] "Moral Epistemology: The Mathematics Analogy." *Noûs*. Vol. 48. 238–55.

Clarke-Doane, Justin. [2015a] "Justification and Explanation in Mathematics and Morality," in Russ Shafer-Landau (ed.), *Oxford Studies in Metaethics, Vol. 10*. New York: Oxford University Press.

Clarke-Doane, Justin. [2015b] "Objectivity in Ethics and Mathematics," in Ben Colburn (ed.), *Proceedings of the Aristotelian Society, The Virtual Issue, No. 3 (Methods in Ethics)*. Available online at http://www.aristoteliansociety.org.uk/pdf/2015_virtual_issue.pdf.

Clarke-Doane, Justin. [2016a] "Debunking and Dispensability," in Neil Sinclair and Uri Leibowitz (eds.), *Explanation in Ethics and Mathematics: Debunking and Dispensability*. Oxford: Oxford University Press.

Clarke-Doane, Justin. [2016b] "What is the Benacerraf Problem?" in Pataut, Fabrice (ed.), *New Perspectives on the Philosophy of Paul Benacerraf: Truth, Objects, Infinity*. Dordrecht: Springer.

Clarke-Doane, Justin. [2019] "Modal Objectivity." Noûs.Vol. 53. 266-295.

Clarke-Doane, Justin. [Forthcoming] "Metaphysical and Absolute Possibility." Synthese (special issue).

Clarke-Doane, Justin. [Manuscript] "Platonic Semantics."

Cohen, G. A. [2000] *If You're an Egalitarian, How Come You're So Rich?* Cambridge, MA: Harvard University Press.

Cohen, Jonathan. [2009] *The Red and the Real.* Oxford: Oxford University Press.

Cohen, Paul.[1971] "Comments on the Foundations of Set Theory," in *Axiomatic Set Theory (Proceedings of Symposia in Pure Mathematics)*, Vol. 13. Providence, RI: American Mathematical Society. 9–15.

Colyvan, Mark. [2007] "Mathematical Recreation versus Mathematical Knowledge," in Leng, Mary, Alexander Paseau, and Michael Potter (eds.), *Mathematical Knowledge*. Oxford: Oxford University Press. 109–22.

Colyvan, Mark. [2015] "Indispensability Arguments in the Philosophy of Mathematics," in *The Stanford Encyclopedia of Philosophy* (Spring Edn), Edward N. Zalta (ed.), https://plato.stanford.edu/archives/spr2015/entries/mathphil-indis/.

Conee, Earl, and Richard Feldman. [1998] "The Generality Problem for Reliabilism." *Philosophical Studies*. Vol. 89. 1–29

Cornwell, Stuart. [1992] "Counterfactuals and the Applications of Mathematics." *Philosophical Studies*. Vol. 66. 73–87.

Craig, William. [1953] "On Axiomatizability within a System." *Journal of Symbolic Logic*. Vol. 18. 30–2.

Crisp, Roger. [2006] *Reason and the Good*. Oxford: Clarendon Press.

Cruz, Joseph, and John Pollock. [1999] *Contemporary Theories of Knowledge (2nd edn)*. Lanham, MD: Rowman & Littlefield.

Cuneo, Terence, and Russ Shafer-Landau. [2014] "The Moral Fixed Points: New Directions for Moral Nonnaturalism." *Philosophical Studies*. Vol. 171. 399–443.

Curry, Oliver Scott, Daniel Austin Mullins, and Harvey Whitehose. [2019] "Is It Good to Cooperate? Testing the Theory of Morality-as-Cooperation in 60 Societies." *Current Anthropology*. Vol. 60. 47–69.

Dahaene, Stanislas. [1997] *The Number Sense: How the Mind Creates Mathematics*. Oxford: Oxford University Press.

Darwin, Charles. [1871] *The Descent of Man*. New York: Appleton.

Das, Ramon. [Forthcoming] "Moral Pluralism and Companions in Guilt." Christopher Cowie and Richard Rowland (eds.), *Companions in Guilt Arguments in Metaethics*. Routledge.

Dasgupta, Shamik. [2018] "Realism and the Absence of Value." *Philosophical Review*. Vol. 127. 279-322.

De Cruz, Helen. [2006] "Why are some Numerical Concepts More Successful than Others? An Evolutionary Perspective on the History of Number Concepts." *Evolution and Human Behavior*. Vol. 27. 306–23

Devlin, Keith. [1981] "Infinite Trees and the Axiom of Constructibility." *Bulletin of the London Mathematical Society*. Vol. 13. 193–206.

Dogramaci, Sinan [2016] "Explaining our Moral Reliability." *Pacific Philosophical Quarterly*. Vol. 98. 71–86.

Dorr, Cian. [2005] "What We Disagree About when We Disagree About Ontology," in Kalderon, Mark Eli (ed.), *Fictionalism in Metaphysics*. Oxford: Oxford University Press. pp. 234–86.

Dorr, Cian. [2008] "There are No Abstract Objects," in Sider, Ted, John Hawthorne, and Dean Zimmerman (eds.), *Contemporary Debates in Metaphysics*. Oxford: Blackwell.

Dreier, James. [2004] "Meta-ethics and the Problem of Creeping Minimalism," in *Ethics, Philosophical Perspectives*. Vol. 18. Oxford: Wiley-Blackwell. 23–44.

Dutilh Novaes, Catarina. [2015] "Conceptual Genealogy for Analytic Philosophy," in J. Bell, A. Cutrofello, P. M. Livingston (eds.), *Beyond the Analytic-Continental Divide: Pluralist Philosophy in the Twenty-First Century* (Routledge Studies in Contemporary Philosophy). New York and Abingdon: Routledge. 75–108.

Dworkin, Ronald. [1996] "Objectivity and Truth: You'd Better Believe It." *Philosophy and Public Affairs*. Vol. 25. 87–139.

Easwaran, Kenny. [2008] "The Role of Axioms in Mathematics." *Erkenntnis*. Vol. 68. 381–91.

Edidin, Aron. [1995] "What Mathematics is About." *Philosophical Studies*. Vol. 78. 1–31.

Eklund, Matti. [2017] *Choosing Normative Concepts*. Oxford: Oxford University Press.

Eklund, Matti. [Forthcoming] "The Normative Pluriverse." *Journal of Ethics and Social Philosophy*.

Ellis, Brian [1990] *Truth and Objectivity*. Oxford: Basil Blackwell.

Enderton, Herbert. [1977] *Elements of Set Theory*. New York: Academic Press.

Enoch, David. [2006] "Agency, Shmagency: Why Normativity Won't Come from What Is Constitutive of Action." *Philosophical Review*. Vol. 115. 169–98.

Enoch, David. [2011] "The Epistemological Challenge to Metanormative Realism: How Best to Understand It, and How to Cope with It." *Philosophical Studies*. Vol. 148. 413–38.

Enoch, David. [2012] *Taking Morality Seriously*. Oxford: Oxford University Press.

Faraci, David. [2019] "Groundwork for an Explanationist Account of Epistemic Coincidence." *Philosophers' Imprint*. Vol. 19. 1–26.

Feferman, Solomon. [1992] "Why a Little Bit Goes a Long Way: Logical Foundations of Scientifically Applicable Mathematics." *Proceedings of the Philosophy of Science Association*. Vol. 2. 442–55.

Ferrier, Edward. [2019] "Against the Iterative Conception of Set." *Philosophical Studies*. Vol. 176. 2681–703.

Field, Hartry. [1980] Science Without Numbers. Princeton: Princeton University Press.

Field, Hartry. [1985] "Can We Dispense with Spacetime?" Repr., with postscript, in Field [1989].

Field, Hartry. [1985] "Can We Dispense with Spacetime?" Repr., with postscript, in Field [1989].

Field, Hartry. [1989] *Realism, Mathematics, and Modality*. Oxford: Blackwell.

Field, Hartry. [1990] "Mathematics and Modality," in Boolos, George (ed.), *Meaning and Method: Essays in Honor of Hilary Putnam*. Cambridge: Cambridge University Press.

Field, Hartry. [1994] "Are our Mathematical and Logical Concepts Highly Indeterminate?" *Midwest Studies in Philosophy*. Vol. 19. 391–429.

Field, Hartry. [1996] "The A Prioricity of Logic." *Proceedings of the Aristotelian Society*. Vol. 96. 359–79.

Field, Hartry. [1998a] "Mathematical Objectivity and Mathematical Objects," in Laurence, Stephen, and Cynthia Macdonald (eds.), *Contemporary Readings in the Foundations of Metaphysics*. Oxford: Blackwell. 387–403.

Field, Hartry. [1998b] "Which Mathematical Undecidables Have Determinate Truth-Values?" in Dales, H. Garth, and Gianluigi Oliveri (ed.), *Truth in Mathematics*. Oxford: Oxford University Press. 291–310.

Field, Hartry. [2005] "Recent Debates about the A Priori," in T. S. Gendler and J. Hawthorne (eds.), *Oxford Studies in Epistemology, Volume 1*. Oxford: Oxford University Press, 69–88.

Field, Hartry. [2009] "Epistemology without Metaphysics." *Philosophical Studies*. Vol. 143. 249–90.

Fine, Kit. [1994] "Essence and Modality," in J. E. Tomberlin (ed.), *Philosophical Perspectives*, Vol. 8. Oxford: Blackwell.

Fine, Kit. [2001] "The Question of Realism." *Philosophers' Imprint*. Vol. 1. 1–30.

Finlay, Stephen. [2010] "Recent Work on Normativity." *Analysis*. Vol. 70. 331–46.

FitzPatrick, William [2008] "Robust Ethical Realism, Non-Naturalism, and Normativity," in Shafer-Landau, Russ (ed.), *Oxford Studies in Metaethics, Vol. 3*. Oxford: Oxford University Press.

FitzPatrick, William [2016] "Morality and Evolutionary Biology," in The Stanford Encyclopedia of Philosophy (Spring Edn). Edward N. Zalta (ed.), https://plato.stanford.edu/archives/spr2016/entries/morality-biology/.

Fletcher, Guy and Michael Ridge. [2014] *Having it Both Ways: Hybrid Theories and Modern Metaethics*. Oxford: Oxford University Press.

Forster, Thomas. [Forthcoming] *The Axioms of Set Theory*. Cambridge: Cambridge University Press. Available online (in preparation): https://www.dpmms.cam.ac.uk/~tf/axiomsofsettheory.pdf. Accessed on July 4, 2018.

Fraenkel, Abraham, Yehoshua Bar-Hillel, and Azriel Levy. [1973] *Foundations of Set Theory (Studies in Logic and the Foundations of Mathematics, Volume 67)*. New York: Elsevier Science Publishers.

Frances, Bryan. [2005] *Scepticism Comes Alive*. New York: Oxford University Press.

Frankena, William. [1973] *Ethics (Foundations of Philosophy Series), 2nd edn*. Englewood Cliffs, N.J: Prentice-Hall. Available online at http://www.ditext.com/frankena/ethics.html.

Franklin, James. [2014] *An Aristotelian Realist Philosophy of Mathematics: Mathematics as the Science of Quantity and Structure*. London: Palgrave Macmillan.

Frege, Gottlob. [1980/1884] *The Foundations of Arithmetic: A Logico-Mathematical Inquiry into the Concept of Number (2nd rev. edn)*. Austin, J. L. (trans.), Evanston, IL: Northwestern University Press.

Friedman, Harvey. [1973] "The Consistency of Classical Set Theory Relative to a Set Theory with Intuitionistic Logic." *Journal of Symbolic Logic*. Vol. 38. 315–19.

Friedman, Harvey. [2000] "Re: FOM: Does Mathematics Need New Axioms?" Post on the *Foundations of Mathematics (FOM)* Listserv. May 25, 2000. Available online at http://www.personal.psu.edu/t20/fom/postings/0005/msg00064.html.

Gaifman, Haim. [2012] "On Ontology and Realism in Mathematics." *Review of Symbolic Logic.* Vol. 5. 480–512.

Gibbard, Alan. [1990] *Wise Choices, Apt Feelings.* Cambridge, MA: Harvard University Press.

Gibbard, Alan. [2003] *Thinking How to Live.* Cambridge: Harvard University Press.

Gibbard, Alan. [2017] "Ethics and Science: Is Plausibility in the Eye of the Beholder?" *Ethical Theory Moral Practice.* Vol. 20. 737–49.

Gibbons, John. [2014] "Knowledge versus Truth," in Littlejohn, Clayton, and John Turri, *Epistemic Norms: New Essays on Action, Belief, and Assertion.* Oxford: Oxford University Press.

Gill, Michael. [2007] "Moral Rationalism vs. Moral Sentimentalism: Is Morality More Like Math or Beauty?" *Philosophy Compass.* Vol. 2. 16–30.

Gill, Michael. [2019] "Morality is Not Like Mathematics: The Weakness of the Math-Moral Analogy." *Southern Journal of Philosophy.* Vol. 57. 194-216

Girle, Rod. [2009] *Modal Logics and Philosophy.* New York: Routledge.

Godel, Kurt. [1947] "What is Cantor's Continuum Problem?" Benacerraf, Paul, and Hilary Putnam (eds.), *Philosophy of Mathematics: Selected Readings (2nd edn).* Cambridge: Cambridge University Press.

Godel, Kurt. [1990/1938] "The Consistency of the Axiom of Choice and the Generalized Continuum Hypothesis," in Feferman, Solomon (ed.), *Godel's Collected Works, Vol. II.* New York: Oxford University Press.

Godel, Kurt. [1990/1947] "Russell's Mathematical Logic," in Feferman, Solomon (ed.), *Godel's Collected Works, Vol. II.* New York: Oxford University Press.

Goldman, Alvin. [1967] "A Causal Theory of Knowing." *Journal of Philosophy.* Vol. 64. 357–72.

Goodman, Nelson. [1955] *Fact, Fiction, and Forecast.* Cambridge, MA: Harvard University Press.

Goodman, Nelson. [1983] *Fact, Fiction, and Forecast (4th edn).* Cambridge, MA: Harvard University Press.

Greene, Joshua. [2013] *Moral Tribes: Emotion, Reason, and the Gap Between Us and Them.* New York: Penguin.

Griffiths, Paul E., and John S. Wilkins [2015] in Sloan, Phillip R., Gerald McKenny, and Kathleen Eggleson (eds.), *Darwin in the Twenty-First Century.* South Bend, IN: University of Notre Dame Press. 201–31.

Hale, Bob. [2013] *Necessary Beings.* Oxford: Oxford University Press.

Hamkins, Joel David. [2011] "The Set-Theoretic Multiverse". arXiv. Available online at https://arxiv.org/abs/1108.4223.

Hamkins, Joel David. [2012] "The Set-Theoretic Multiverse." (rev. version) *Review of Symbolic Logic.* Vol. 5. 416–49.

Hamkins, Joel David. [2014] "Re: Is there any Research on Set Theory without Extensionality Axiom?" Post on *MathOverflow.* May 27, 2014. Available online at https://mathoverflow.net/questions/168287/is-there-any-research-on-set-theory-without-extensionality-axiom.

Hamkins, Joel David. [2015] "Is the Dream Solution of the Continuum Hypothesis Attainable?" *Notre Dame Journal of Formal Logic*. Vol. 56. 135–45.

Hare, R. M. [1997] *Sorting Out Ethics*. Oxford: Clarendon Press.

Harman, Gilbert. [1977] *The Nature of Morality: An Introduction to Ethics*. New York: Oxford University Press.

Harman, Gilbert. [1986] "Moral Explanations of Natural Facts – Can Moral Claims be Tested Against Realty?" *Southern Journal of Philosophy*. Vol. 24. 57–68.

Harman, Gilbert. [2002] "Moral Realism is Moral Relativism." Manuscript. Available online at https://www.princeton.edu/~harman/Papers/Relativism_Realism.pdf.

Hart. W. D. [1996] "Introduction," in *Philosophy of Mathematics*. Oxford: Oxford University Press.

Hellman, Geoffrey. [1989] *Mathematics without Numbers*. Oxford: Oxford University Press.

Herman, Barbara. [2007] *Moral Literacy*. Cambridge, MA: Harvard University Press.

Hilbert, David. [1983/1936] "On the Infinite," in Benacerraf, Paul, and Hilary Putnam (eds.), *Philosophy of Mathematics: Selected Readings (2nd edn)*. Cambridge: Cambridge University Press.

Hirvela, J. (2017). "Is it Safe to Disagree?" *Ratio*. Vol. 30. 305–21.

Hofweber, Thomas. [2017] "Are There Ineffable Aspects of Reality?" in Bennett, Karen, and Dean Zimmerman (eds.), *Oxford Studies in Metaphysics, Vol. 2*. Oxford: Oxford University Press. 155–206.

Horgan, Terence, and Mark Timmons. [1992] "Troubles on Moral Twin Earth: Moral Queerness Revived." *Synthese*. Vol. 92. 221–60.

Horgan, Terence. [1987] "Discussion: Science Nominalized Properly." *Philosophy of Science*. Vol. 54. 281–2.

Horwich, Paul. [1998] *Truth (2nd edn)*. Oxford: Oxford University Press.

Huemer, Michael. [2005] *Ethical Intuitionism*. New York: Palgrave Macmillan.

Jackson, Frank. [1998] *From Metaphysics to Ethics: A Defense of Conceptual Analysis*. Oxford: Clarendon Press.

Jeshion, Robin. [2011] "Experience as a Natural Kind: Reflections on Albert Casullo's *A Priori Justification*," in Shaffer, Michael, and Michael Veber (eds.), *What Place for the A Priori?* La Salle, IL: Open Court Press.

Jensen, Ronald. [1995] "Inner Models and Large Cardinals." *Bulletin of Symbolic Logic*. Vol. 1. 393—407.

Jonas, Silvia. [2017] "Access Problems and Explanatory Overkill." *Philosophical Studies*. Vol. 174. 2731–42.

Jonas, Silvia. [Forthcoming] "Mathematical and Moral Disagreement." *Philosophical Quarterly*.

Joyce, Richard. [2001] *The Myth of Morality*. Cambridge: Cambridge University Press.

Joyce, Richard. [2007] *The Evolution of Morality*. Cambridge: MIT Press.

Joyce, Richard. [2008] "Precis of The Evolution of Morality." *Philosophy and Phenomenological Research*. Vol. 77. 213–18.

Joyce, Richard. [2016] "Evolution, Truth-Tracking, and Moral Skepticism," in *Essays in Moral Skepticism*. Oxford: Oxford University Press.

Jubien, Michael. [1997] *Contemporary Metaphysics: An Introduction*. Oxford: Blackwell.

Jubien, Michael. [2006] "Property-Theoretic Foundations of Mathematics," in Jacquette, Dale (ed.), *A Companion to Philosophical Logic*. Oxford: Blackwell.

Kahane, Guy. [2011] "Evolutionary Debunking Arguments." *Noûs*. Vol. 45. 103–25.

Kaspar, David. [2015] "Explaining Intuitionism." *Reasons Papers*. Vol. 37. 47–66.

Kelly, Thomas, and Sarah McGrath. [2010] "Is Reflective Equilibrium Enough?" *Philosophical Perspectives*. Vol. 1. 325–259.

Kilmister, Clive W. [1980] "Zeno, Aristotle, Weyl and Shuard: Two-and-a-Half Millenia of Worries over Number." *Mathematical Gazette*. Vol. 64. 149–58.

Kitcher, Philip. [1985] *The Nature of Mathematical Knowledge*. Oxford: Oxford University Press.

Kitcher, Philip. [2011] "Philosophy Inside Out." *Metaphilosophy*. Vol. 42. 248–60.

Klenk, Michael. [Manuscript] "Modal Security and Moral Objectivity."

Kment, Boris. [2014] *Modality and Explanatory Reasoning*. Oxford: Oxford University Press.

Koellner, Peter. [2014] "On the Question of Absolute Undecidability." *Philosophia Mathematica*. Vol. 14. 153–88.

Koellner, Peter, Joan Bagaria, and Hugh Woodin. [2017] "Large Cardinals Beyond Choice." *XXVI incontro dell'Associazione Italiana di Logica e sue Applicazioni (Meeting of the Italian Association of Logic and its Applications)*. September 25–28, 2017. Slides available online at https://events.math.unipd.it/aila2017/sites/default/files/BAGARIA.pdf.

Korsgaard, Christine. [1996] *The Sources of Normativity*. New York: Cambridge University Press.

Kriesel, George. [1967] "Observations on Popular Discussions of the Foundations of Mathematics," in Scott, Dana. (ed.) *Axiomatic Set Theory (Proceedings of Symposia in Pure Mathematics, V. XIII, Part I)*. Providence, RI: American Mathematical Association.

Kripke, Saul. [1963] "Semantical Considerations on Modal Logic." *Acta Philosophica Fennica*. Vol. 16. 83–94.

Kripke, Saul. [1980] *Naming and Necessity*. Cambridge: Harvard University Press.

Kripke, Saul. [2011] "Vacuous Names and Fictional Entities," in *Philosophical Troubles: Collected Papers, Vol. 1*. New York: Oxford University Press.

Kuhn, Thomas. [1962] *The Structure of Scientific Revolutions*. Chicago: University of Chicago Press.

Kunen, Kenneth. [1980] *Set Theory: An Introduction to Independence Proofs*. Amsterdam: Elsevier.

Lakatos, Imre. [1976] "A Renaissance of Empiricism in the Recent Philosophy of Mathematics." *British Journal for the Philosophy of Science*. Vol. 27. 201–23.

Lear, Jonathan. [1983] "Ethics, Mathematics, and Relativism." *Mind*. Vol. 92. 38-60.

Leibowitz, Uri D., and Neil Sinclair (eds.). [2016] *Explanation in Ethics and Mathematics: Debunking and Dispensability*. Oxford: Oxford University Press.

Leiter, Brian. [2001] "Moral Facts and Best Explanations." *Social Philosophy and Policy*. Vol. 18. 79–101.

Leiter, Brian. [2009] "Moral Skepticism and Moral Disagreement in Nietzsche." *Public Law Working Paper no. 257*. University of Chicago. Available online at http://ssrn.com/abstractp1315061.

Leiter, Brian. [2010] "Moral Skepticism and Moral Disagreement: Developing an Argument from Nietzsche." *On the Human (National Humanities Center).* March 25, 2010. Available online at https://nationalhumanitiescenter.org/on-the-human/2010/03/moral-skepticism-and-moral-disagreement-developing-an-argument-from-nietzsche/.

Leng, Mary. [2007] "What's There to Know?" in Leng, M., Paseau, A., and Potter, M. (eds.), *Mathematical knowledge.* Oxford: Oxford University Press.

Leng, Mary. [2009] ""Algebraic" Approaches to Mathematics," in Otavio Bueno and Oystein Linnebo, *New Waves in the Philosophy of Mathematics.* New York: Palgrave Macmillan.

Leng, Mary. [2010] *Mathematics and Reality.* Oxford: Oxford University Press.

Lewis, David. [1983] "New Work for a Theory of Universals." *Australasian Journal of Philosophy.* Vol. 61. 343–77.

Lewis, David. [1986] *On the Plurality of Worlds.* Oxford: Blackwell.

Liggins, David. [2006] "Is there a Good Epistemological Argument against Platonism?" *Analysis.* Vol. 66. 135–41.

Liggins, David. [2010] "Epistemological Objections to Platonism." *Philosophy Compass.* Vol. 5. 67–77.

Liggins, David. [2014] "Grounding, Explanation, and Multiple Realization in Mathematics and Ethics," in Sinclair, Neil, and Uri Leibowitz (eds.), E*xplanation in Ethics and Mathematics: Debunking and Dispensability.* Oxford: Oxford University Press.

Lillehammer, Hallvard. [2007] *Companions in Guilt: Arguments for Ethical Objectivity.* London: Palgrave MacMillan.

Linnebo, Øystein. [2006] "Epistemological Challenges to Mathematical Platonism." *Philosophical Studies.* Vol. 129. 545–74.

Linsky, Bernard, and Edward Zalta. [1995] "Naturalized Platonism versus Platonism Naturalized." *Journal of Philosophy.* Vol. 92. 525—55.

Locke, Dustin, and Daniel Korman. [Forthcoming] "Against Minimalist Responses to Debunking Arguments," in Shafer-Landau, Russ (ed.), *Oxford Studies in Metaethics.* Draft available online at http://www.marcsandersfoundation.org/wp-content/uploads/KormanandLocke.pdf.

Lowe, E. J. [1993]: "Are the Natural Numbers Individuals or Sorts?" *Analysis.* Vol. 53. 147–54.

Lowe, E. J. [1995]: "The Metaphysics of Abstract Objects." *Journal of Philosophy.* Vol. 92. 509–24.

Lowe, E. J. [2012] "What is the Source of Our Knowledge of Modal Truths?" *Mind.* Vol. 121. 919—50.

Lyon, Aidan and Mark Colyvan. [2008] "The Explanatory Power of Phase Spaces." *Philosophia Mathematica.* Vol. 16. 227–43.

Maddy, Penelope. [1988a] "Believing the Axioms: I." *Journal of Symbolic Logic.* Vol. 53. pp. 481–511.

Maddy, Penelope. [1988b] "Believing the Axioms: II." *Journal of Symbolic Logic.* Vol. 53. pp. 736–64.

Maddy, Penelope. [1990] *Realism in Mathematics.* Oxford: Oxford University Press.

Maddy, Penelope. [1997] *Naturalism in Mathematics.* Oxford: Clarendon Press.

Maddy, Penelope. [2008] "How Applied Mathematics Became Pure." *Review of Symbolic Logic*. Vol. 1. 16–41.

Majors, Brad. [2007] "Moral Explanation." *Philosophy Compass*. Vol. 2. 1–15.

Malament, David. [1982] "Review of *Science without Numbers*." *Journal of Philosophy*. Vol. 79. 523–34.

Marshall, Colin. [2018] *Compassionate Moral Realism*. Oxford: Oxford University Press.

Martin, D. A. [1976] "Hilbert's First Problem: The Continuum Hypothesis," in Browder, Felix (ed.), *Mathematical Developments Arising from Hilbert Problems (Proceedings of Symposia in Pure Mathematics. Vol. 28)*. Providence, RI: American Mathematical Society.

Martin, D. A. [1998] "Mathematical Evidence," in Dales, H. G., and G. Oliveri (eds.), *Truth in Mathematics*. Oxford: Clarendon Press.

Martin, M. G. F. [1997] "The Reality of Appearances," in Sainsbury, Michael (ed.), *Thought and Ontology*. Milan: Franco Angeli. 77–96.

Mayberry, John. [2000] *The Foundations of Mathematics in the Theory of Sets*. Cambridge: Cambridge University Press.

McDowell, John. [2008] "The Disjunctive Conception of Experience as Material for a Transcendental Argument", in Macpherson, Fiona, and Adrian Haddock (eds.), *Disjunctivism: Perception, Action, Knowledge*. Oxford: Oxford University Press. 376–89.

McGrath, Matthew. [2018] "Looks and Perception Justification." *Philosophy and Phenomenological Research*. Vol. 96. 110–33.

McGrath, Sarah. [2007] "Moral Disagreement and Moral Expertise." *Oxford Studies in Metaethics*. Vol. 3. 87–108.

McGrath, Sarah. [2010] "Moral Knowledge and Experience." *Oxford Studies in Metaethics*. Vol. 6. 107–27.

Mackie, J. L. [1977] *Ethics: Inventing Right and Wrong*. Harmondsworth: Penguin.

McPherson, Tristram, and David Enoch. [2017] "What do you mean 'This isn't the Question'?" *Canadian Journal of Philosophy*. Vol. 47. 820–40.

McPherson, Tristram,. [2018] "Naturalistic Moral Realism, Rationalism, and Non-Fundamental Epistemology," in Jones, Karen, and Francois Schroeter (eds.), *The Many Moral Rationalisms*. Oxford: Oxford University Press.

McPherson, Tristram,. [Forthcoming b] "Authoritatively Normative Concepts." Shafer-Landau (ed.), Oxford Studies in Metaethics. Available online at https://philpapers.org/archive/MCPANC.pdf.

Melia, Joseph. [2000] "Weaseling Away the Indispensability Argument." *Mind*. Vol. 109. 455–79.

Melia, Joseph. [2014] *Modality*. New York: Routledge.

Merricks, Trenton. [2001] *Objects and Persons*. Oxford: Oxford University Press.

Mill, John Stuart. [1863] *Utilitarianism*. London: Parker, Son, and Bourne, West Strand. Available online at https://archive.org/stream/a592840000milluoft#page/n3/mode/2up.

Mill, John Stuart. [2009/1882] *A System Of Logic, Ratiocinative And Inductive* (8th edn). New York: Harper & Brothers. Available online at https://www.gutenberg.org/files/27942/27942-pdf.pdf.

Mogensen, Andreas. [2016] "Contingency Anxiety and the Epistemology of Disagreement." *Pacific Philosophical Quarterly*. Vol. 97.

Moore, G.E. [1903] Principia Ethica. Available online at: <http://fair-use.org/g-e-moore/principia-ethica>

Mulvey, John (ed.). [1981] *The Nature of Matter*. Oxford: Oxford University Press.

Nagel, Thomas. [1986] *The View from Nowhere*. Oxford: Oxford University Press.

Nagel, Thomas. [1997] *The Last Word*. New York: Oxford University Press.

Nietzsche, Friedrich. [1887] On the Genealogy of Morality. Available online at:<https://archive.org/details/genealogyofmoral00nietuoft/page/n6>

Nelson, Edward. [1986] *Predicative Arithmetic (Mathematical Notes. No. 32)*. Princeton, NJ: Princeton University Press.

Nelson, Edward. [2013] "Re: Illustrating Edward Nelson's Worldview with Nonstandard Models of Arithmetic." Post on *MathOverflow*. October 31, 2013. Available online at https://mathoverflow.net/questions/142669/illustrating -edward-nelsons-worldview-with-nonstandard-models-of-arithmetic.

Nozick, Robert. [1981] *Philosophical Explorations*. Cambridge: Cambridge University Press.

Papineau, David. [2019] "Knowledge is Crude." Aeon. Available online at https:// aeon.co/essays/knowledge-is-a-stone-age-concept-were-better-off-without-it.

Parfit, Derek. [1984] *Reasons and Persons*. Oxford: Oxford University Press.

Parfit, Derek. [2011] *On What Matters: Volume 2*. Oxford: Oxford University Press.

Parsons, Charles. [2009] *Mathematical Thought and Its Objects*. Cambridge: Cambridge University Press.

Paseau, Alexander. [2009] "Reducing Arithmetic to Set Theory," in Bueno, Otávio, and Øystein Linnebo (eds.), *New Waves in Philosophy of Mathematics. New Waves in Philosophy*. London: Palgrave Macmillan.

Peacocke, Christopher. [1999] *Being Known*. Oxford: Oxford University Press.

Peacocke, Christopher. [2004] *The Realm of Reason*. Oxford: Oxford University Press.

Pigliucci, Massimo. [2018] "Is There a Universal Morality?" The Evolution Institute. Available online at https://evolution-institute.org/is-there-a-universal-morality/.

Pinker, Steven. [2002] The Blank Slate: The Modern Denial of Human Nature. London: Penguin Books.

Pinter, Charles. [2014/1971] A Book of Set Theory. Mineola, NY: Dover.

Pinker, Steven. [2008] "The Moral Instinct." The Stone *(New York Times)*. January 13, 2008. Available online at https://www.nytimes.com/2008/01/13/magazine/ 13Psychology-t.html.

Pollock, John. [1986] *Contemporary Theories of Knowledge*. Totowa, NJ: Rowman & Littlefield.

Potter, Michael. [2004] *Set Theory and Its Philosophy: A Critical Introduction*. Oxford: Oxford University Press.

Priest, Graham. [2012] "Mathematical Pluralism." *Logic Journal of the IGPL*. Vol. 2. 4–13.

Pritchard, Duncan. [2009] "Safety-Based Epistemology: Whither Now?" *Journal of Philosophical Research*. Vol. 34. 33–45.

Pryor, James. [2000] "The Skeptic and the Dogmatist." *Noûs*. Vol. 34. 517–49.

Pudlák, Pavel. [2013] Logical Foundations of Mathematics and Computational Complexity Theory: A Gentle Introduction. New York: Springer.

Putnam, Hilary. [1965] "Craig's Theorem." Journal of Philosophy. Vol. 62. 251–60.

Putnam, Hilary. [1967] "Mathematics Without Foundations." Journal of Philosophy. Vol. 64. 5–22.

Putnam, Hilary. [1971] Philosophy of Logic (Essays in Philosophy). New York: Harper & Row.

Putnam, Hilary. [1979/1994] "Philosophy of Mathematics: Why Nothing Works," in Putnam, Hilary, Words and Life. Cambridge: Harvard University Press.

Putnam, Hilary. [1980] "Models and Reality." Journal of Symbolic Logic. Vol. 45. 464–82.

Putnam, Hilary. [2004] Ethics Without Ontology. Cambridge: Harvard University Press.

Putnam, Hilary. [2012] "Indispensability Arguments in the Philosophy of Mathematics," in De Caro, Mario, and David Macarthur (eds.), Philosophy in the Age of Science: Physics, Mathematics, Skepticism. Cambridge: Harvard University Press.

Putnam, Hilary. [2015] "Intellectual Autobiography," in E. Auxier, Randall, Douglas R. Anderson, and Lewis Edwin Hahn (eds.), The Philosophy of Hilary Putnam (Library of Living Philosophers, Vol. XXXIV). Chicago, IL: Open Court.

Quine, W. V. O. [1937] "New Foundations for Mathematical Logic." American Mathematical Monthly. Vol. 44. 70–80.

Quine, W. V. O. [1948] "On What There Is." Review of Metaphysics. Vol. 2. 21–38.

Quine, W. V. O. [1951a] "Ontology and Ideology." Philosophical Studies. Vol. 2. 11–15.

Quine, W. V. O. [1951b] "Two Dogmas of Empiricism." Philosophical Review. Vol. 60. 20–43.

Quine, W. V. O. [1953] "Three Grades of Modal Involvement." Proceedings of the XIth International Congress of Philosophy. Vol. 14. 65–81.

Quine, W. V. O. [1964] "Ontological Reduction and the World of Numbers." Journal of Philosophy. Vol. 61. 209–16.

Quine, W. V. O. [1969] Set Theory and Its Logic (rev. edn). Cambridge: Harvard University Press.

Quine, W. V. O. [1986a] "Reply to Charles Parsons," in Hahn, Lewis Edwin, and Paul Arthur Schilp (eds.). The Philosophy of W. V. Quine (Library of Living Philosophers, Vol. 18). La Salle, IL: Open Court.

Quine, W. V. O. [1986b] Philosophy of Logic (2nd edn). Cambridge, MA: Harvard University Press.

Quine, W. V. O. [1990] Pursuit of Truth. Cambridge: Harvard University Press.

Pantsar, Markus. [2014] "An Empirically Feasible Approach to the Epistemology of Arithmetic." Synthese. Vol. 191. 4201—4229.

Rachels, James. (ed.) [1998] "Introduction," in Ethical Theory 1: The Question of Objectivity (Oxford Readings in Philosophy). Oxford: Oxford University Press.

Railton, Peter. [1986] "Moral Realism." The Philosophical Review. Vol. 95. 163–207.

Railton, Peter. [2003] "Noncogntivism about Rationality: Benefits, Costs, and an Alternative," in his Facts, Values, and Norms: Essays Toward a Morality of Consequence. Cambridge: Cambridge University Press.

Railton, Peter. [2006] "Moral Factualism," in Drier, James (ed.), *Contemporary Debates in Moral Theory*. Oxford: Blackwell.

Rawls, John. [1971] *A Theory of Justice*. Cambridge, MA: Harvard University Press.

Rawls, John. [1974] "The Independence of Moral Theory." *Proceedings and Addresses of the American Philosophical Association*. Vol. 47. 5–22.

Rayo, Agustin. [2013] *The Construction of Logical Space*. Oxford: Oxford University Press.

Reichenbach, Hans. [1956/1927] *The Philosophy of Space and Time*. Freund, John, and Maria Reichenbach (trans.), New York: Dover.

Reicher, Maria. [2019] "Nonexistent Objects," in The Stanford Encyclopedia of Philosophy (Spring Edn), Edward N. Zalta (ed.), https://plato.stanford.edu/entries/nonexistent-objects/.

Reid, Thomas. [1983, 1788] "Of Systems of Morals," in his *Philosophical Works* (facsimile of the 1895 edn). New York: George Holmes Publishers.

Relaford-Doyle, Josephine and Rafael Núñez. [2018] "Beyond Peano: Looking into the Unnaturalness of the Natural Numbers," in Bangu, Sorin (ed.), *Naturalizing Logico-Mathematical Knowledge Approaches from Philosophy, Psychology and Cognitive Science*. New York: Routledge.

Restall, Greg, and J. C. Beall. [2006] *Logical Pluralism*. Oxford: Oxford University Press.

Ridge, Michael [2007] "Anti-Reductionism and Supervenience." *Journal of Moral Philosophy*. Vol. 4. 330–48.

Rieger, Adam. [2011] "Paradox, ZF, and the Axiom of Foundation," in DeVidi, David, Hallet, Michael, and Peter Clark (eds.). *Logic, Mathematics, Philosophy, Vintage Enthusiasms: Essays in Honour of John L. Bell (The Western Ontario Series in Philosophy of Science)*. New York: Springer.

Roberts, Debbie. [2016] "Explanatory Indispensability Arguments in Metaethics and the Philosophy of Mathematics," in Uri D. Leibowitz and Neil Sinclair (eds.), *Explanation in Ethics and Mathematics: Debunking and Dispensability*. Oxford: Oxford University Press.

Rosen, Gideon. [1994] "Objectivity and Modern Idealism: What is the Question?" in O'Leary-Hawthorne, John, and Michaelis Michael (eds.), *Philosophy in Mind*. Dordrecht: Kluwer Academic Publishers. 277–319.

Rosen, Gideon. [2001] "Nominalism, Naturalism, and Epistemic Relativism." *Philosophical Perspectives*. Vol. 15. 69–91.

Rosen, Gideon. [2014, Manuscript] "What is Normative Necessity?" Available online at www.academia.edu/9159728/Normative_Necessity.

Rosenberg, Alexander. [2015] "Can Moral Disputes be Resolved?" *The Stone (New York Times)*. July 13, 2015. Available online at https://opinionator.blogs.nytimes.com/2015/07/13/can-moral-disputes-be-resolved/.

Rosenberg, Alexander. [2018] Interview for *What Is It Like to be a Philosopher?* February 23, 2018. Available online at http://www.whatisitliketobeaphilosopher.com/#/alex-rosenberg/.

Ross, W. D. [1930] *The Right and the Good*. Available online at http://www.ditext.com/ross/right.html>.

Rovane, Carol. [2013] *The Metaphysics and Ethics of Relativism*. Cambridge, MA: Harvard University Press.

Roychoudhuri, Onnesha. [2007] "Our Rosy Future According to Freedman Dyson." Interview for Salon magazine. Available online at https://www.salon.com/2007/09/29/freeman_dyson/.

Ruse, Michael. [1986] *Taking Darwin Seriously*. Oxford: Blackwell.

Russell, Gillian. [2011] Truth in Virtue of Meaning: A Defense of the Analytic/Synthetic Distinction. Oxford: Oxford University Press.

Russell, Bertrand. [1905] "On Denoting." *Mind*. Vol. 14. 479–93.

Russell, Bertrand. [1918] "The Philosophy of Logical Atomism." *The Monist*. Vol. XXVIII. 495–526.

Russell, Bertrand. [1948] *Human Knowledge: Its Scope and Limits*. London: George Allen & Unwin Ltd.

Russell, Bertrand. [1973/1907] "The Regressive Method for Discovering the Premises of Mathematics," in Lackey, Douglas (ed.). *Essays in Analysis*, by Bertrand Russell. London: Allen & Unwin Ltd.

Saatsi, Juha. [2016] "On the 'Indispensable Explanatory Role' of Mathematics." *Mind*. Vol. 125. 1045–70.

Sartre, Jean-Paul. [1946] "Existentialism is Humanism." Available online at https://www.marxists.org/reference/archive/sartre/works/exist/sartre.htm.

Sayre-McCord. [1988] "Moral Theory and Explanatory Impotence." *Midwest Studies in Philosophy*. Vol. 12. 433–57.

Scanlon, T.M. [2014] *Being Realistic about Reasons*. New York, NY: Oxford University Press.

Schafer, Karl. [2017] "Review of Sinclair, Neil and Uri Leibowitz (eds.), Explanation in Ethics and Mathematics: Debunking and Dispensability." *Notre Dame Philosophical Reviews*. May 25, 2017. Available online at https://ndpr.nd.edu/news/explanation-in-ethics-and-mathematics-debunking-and-dispensability/.

Schechter, Joshua. [2006] *Two Challenges to the Objectivity of Logic*. PhD dissertation. New York University, College of Arts and Sciences.

Schechter, Joshua. [2010] "The Reliability Challenge and the Epistemology of Logic." *Philosophical Perspectives*. Vol. 24. 437–64.

Schechter, Joshua. [2013] "Could Evolution Explain our Reliability about Logic?" in Hawthorne, John, and Tamar Szabò (eds.). *Oxford Studies in Epistemology, Vol. 4*. 214–39. Oxford: Oxford University Press.

Schechter, Joshua. [2018] "Is There a Reliability Challenge for Logic?" *Philosophical Issues*. Vol. 28. 325 – 347.

Schiffer, Stephen. [2003] *The Things We Mean*. Oxford: Oxford University Press.

Shoenfield, Joseph. [1977] "The Axioms of Set Theory," in Barwise, John (ed.), *Handbook of Mathematical Logic*. Amsterdam: Elsevier. 321–44.

Scott, Dana. [1961] "Moral on the Axiom of Extensionality," in Bar-Hillel, Y., E.I.J. Poznanski, M.O. Rabin, and A. Robinson (eds.), *Essays on the Foundations of Mathematics, Dedicated to A.A. Fraenkel on his Seventieth Anniversary*. Jerusalem: Magnes Press.

Schroeder, Mark. [2010] *Being For: Evaluating the Semantic Program for Expressivism*. New York: Oxford University Press.

Schroeter, Laura, and Francois Schroeter. [2013] "Co-reference without Convergence?" *Philosophers' Imprint*. Vol. 13. 1–34. Available online at https://quod.lib.umich.edu/cgi/p/pod/dod-idx/normative-realism-co-reference-without-convergence.pdf?c=phimp;idno=3521354.0013.013;format=pdf.

Segal, Aaron. [2019] "Pythagoreanism: A Number of Theories." *Philosophers' Imprint*. Vol. 19. 1—19.

Sellars, Wilfrid. [2007/1962] "Philosophy and the Scientific Image of Man," in Brandom, Robert, and Kevin Scharp (eds.), *In the Space of Reasons: Selected Essays from Wilfrid Sellars*. Cambridge: Harvard University Press.

Shafer-Landau. [2006] *Moral Realism: A Defense*. Oxford: Oxford University Press.

Shafer-Landau. [2009] "A Defense of Categorical Reasons." *Proceedings of the Aristotelian Society*, Vol. 109. 189–206.

Shapiro, Stewart. [1995] "Modality and Ontology". *Mind*. Vol. 102. 455–81.

Shapiro, Stewart. [2000] *Thinking about Mathematics: Philosophy of Mathematics*. Oxford: Oxford University Press.

Shapiro, Stewart. [2009] "We Hold These Truths to be Self-Evident: But What Do We Mean By That?" *Review of Symbolic Logic*. Vol. 2. 175–207.

Shapiro, Stewart. [2014] *Varieties of Logic*. Oxford: Oxford University Press.

Sider, Ted. [2011] *Writing the Book of the World*. New York: Oxford University Press.

Siegel, Ethan. [2018] "The Biggest Myth in Quantum Physics." Science *(Forbes Magazine)*. February 7, 2018. Available online at https://www.forbes.com/sites/startswithabang/2018/02/07/the-biggest-myth-in-quantum-physics/#402a354f53fa.

Siegel, Susanna. [2016] "The Contents of Perception," in The Stanford Encyclopedia of Philosophy (Winter Edn), Edward N. Zalta (ed.), https://plato.stanford.edu/archives/win2016/entries/perception-contents/.

Silins, Nico. [2011] "Seeing Through the 'Veil of Perception'." *Mind*. Vol. 120. 329–67.

Sinclair, Neil. [2007] "Expressivism and the Practicality of Normative Convictions." *Journal of Value Inquiry*. Vol. 41. 201–20.

Singer, Peter. [1994] "Introduction," in Singer, Peter (ed.), *Ethics*. Oxford: Oxford University Press.

Sinnott-Armstrong, Walter. [2006] *Moral Skepticisms*. Oxford: Oxford University Press.

Sinnott-Armstrong, Walter. [2018] "AI Alignment Podcast: On Becoming a Moral Realist with Peter Singer." *Future of Life Institute* (interview by Lucus Perry). Posted October 18, 2018. Available online at https://futureoflife.org/2018/10/18/on-becoming-a-moral-realist-peter-singer/.

Smith, Peter. [2013] *An Introduction to Godel's Theorems (2nd edn)*. Cambridge: Cambridge University Press.

Sober, Elliott. [1984] *The Nature of Selection*. Cambridge, MA: MIT Press.

Sober, Elliott. [1993] "Mathematics and Indispensability." *Philosophical Review*. Vol. 102. 35–57.

Stalnaker, Robert. [1996] "On What Possible Worlds Could Not Be," in *Ways a World Might Be: Metaphysical and Anti-Metaphysical Essays*. Oxford: Oxford University Press. 40–54.

Steinhart, Eric. [2002] "Why Numbers are Sets." *Synthese*. Vol. 133. 343–61.

Street, Sharon. [2006] "A Darwinian Dilemma for Realist Theories of Value." Philosophical Studies. Vol. 127. 109—166.

Street, Sharon. [2008] "Reply to Copp: Naturalism, Normativity, and the Varieties of Realism Worth Worrying About." *Philosophical Issues*. Vol. 18. 207–28.

Street, Sharon. [2011] "Mind-Independence without Mystery: Why Quasi-Realists Can't Have it Both Ways." *Oxford Studies in Metaethics*, Vol. 6. 1–32.

Street, Sharon. [2016] "Objectivity and Truth: You'd Better Rethink It," in Shafer-Landua (ed.), *Oxford Studies in Metaethics, Vol. 11*. New York: Oxford University Press.

Streumer, Bart. [2013] "Can We Believe the Error Theory?" *Journal of Philosophy*. Vol. 110. 194–212.

Strohminger, Margot and Juhani Yli-Vakkuri. [2017] "The Epistemology of Modality." Analysis. Vol. 77. 825-838.

Sturgeon, Nicholas. [1984] "Moral Explanations," in Copp, David, and David Zimmerman (eds.), *Morality, Reason, and Truth*. Totowa, NJ: Rowman & Allanheld. 49–78.

Sturgeon, Nicholas. [1986] "Harman on Moral Explanations of Natural Facts." *Southern Journal of Philosophy* (Supplement). Vol. 24. 69–78.

Sturgeon, Nicholas. [1991] "Contents and Causes: A Reply to Blackburn." *Philosophical Studies*. Vol. 61. 19–37.

Sturgeon, Nicholas. [2006] "Moral Explanations Defended," in James Dreier (ed.), *Contemporary Debates in Moral Theory*. Oxford: Blackwell. 241–62.

Tegmark, Max. [2014] *Our Mathematical Universe*. New York: Random House.

Tersman, Folke. [2016] "Explaining the Reliability of Our Moral Beliefs," in Sinclair, Neil, and Uri Leibowitz (eds.), *Explanation in Ethics and Mathematics: Debunking and Dispensability*. Oxford: Oxford University Press.

Thomson, Judith Jarvis, and Gilbert Harman. [1996] *Moral Relativism and Moral Objectivity*. Oxford: Blackwell.

Unger, Peter. [2014] *Empty Ideas: A Critique of Analytic Philosophy*. Oxford: Oxford University Press.

Van Inwagen, Peter. [1990] *Material Beings*. Ithaca, NY: Cornell University Press.

Van Roojen, Mark [2006] "Knowing Enough to Disagree: A New Response to the Moral Twin Earth Argument," in Shafer-Landau (ed.), *Oxford Studies in Metaethics, Vol. 1*. Oxford: Oxford University Press.

Varzi, Achille. [Manuscript] "Counterpart Theory without Modal Realism."

Vavova, Katia. [2018] "Irrelevant Influences." *Philosophy and Phenomenological Research*. Vol. 96. 134–52

Vogt, Katja. [2016] "Ancient Skepticism", in The Stanford Encyclopedia of Philosophy (Winter Edn), Edward N. Zalta (ed.), https://plato.stanford.edu/archives/win2016/entries/skepticism-ancient/.

Warren, Jared. [2015] "Conventionalism, Consistency, and Consistency Sentences." *Synthese*. Vol. 192. 1351–71.

Warren, Jared. [2017] "Epistemology versus Non-Causal Realism." *Synthese*. Vol. 194. 1643–62.

Wedgwood, Ralph. [2001] "Conceptual Role Semantics for Moral Terms." *Philosophical Review*. Vol. 110: 1–30.

Wedgwood, Ralph. [2007] *The Nature of Normativity*. Oxford: Clarendon Press.

Werner, Preston. [2018] "Why Conceptual Competence Won't Help the Non-Naturalist Epistemologist." *Canadian Journal of Philosophy*. Vol. 48. 616–37.

Weyl, Hermann. [1918] *Das Kontinuum: Kritische Untersuchungen über die Grundlagen der Analysis*. Leipzig: Veit.

White, Roger. [2010] "You Just Believe that Because." *Philosophical Perspectives*. Vol. 24. 573–615.

Whitehead, Alfred North, and Bertrand Russell [1997] *Principia Mathematica to *56* (Cambridge Mathematical Library). Cambridge: Cambridge University Press.

Williamson, Timothy. [2000] Knowledge and Its Limits. Oxford: Oxford University Press.

Williamson, Timothy. [2012] "Logic and Neutrality." New York Times (The Stone). May 13, 2012. Available online at <https://opinionator.blogs.nytimes.com/2012/05/13/logic-and-neutrality/>

Williamson, Timothy. [2013] "Very Improbable Knowing." Vol. 79. Erkenntnis. 971–999.

Williamson, Timothy. [2016] "Modal Science." *Canadian Journal of Philosophy*. Vol. 46. 453–92.

Williamson, Timothy. [2017] "Counterpossibles in Semantics and Metaphysics," *Argumenta*. Vol. 2. 195-226. Available online at https://www.argumenta.org/wp-content/uploads/2017/06/2-Argumenta-22-Timothy-Williamson-Counterpossibles-in-Semantics-and-Metaphysics.pdf

Wolterstorff, Nicholas. [1970] *On Universals*. Chicago: University of Chicago Press.

Wolterstorff, Nicholas. [1979] "Characters and their Names." *Poetics*. Vol. 8. 101–27.

Woodin, Hugh. [2001] "The Continuum Hypothesis I." *Notices of the American Mathematical Society*. Vol. 48. 567–576.

Woods, Jack. [2018] "Morality, Mathematics, and Self-Effacement." *Noûs*. Vol. 52. 47–68.

Wright, Crispin. [1994] *Truth and Objectivity*. Cambridge: Harvard University Press.

Wright, Crispin, and Bob Hale. [2002] "Benacerraf's Dilemma Revisited." *European Journal of Philosophy*. Vol. 10. 101–29.

Zach, Richard. [Forthcoming] "Rumfitt on Truth-Grounds, Negation, and Vagueness." *Philosophical Studies*. Available online: <https://philarchive.org/archive/ZACROT-3>

Zielberger, Doron. [2004] "'Real' Analysis is a Degenerate Case of Discrete Analysis", in Aulbach, Bernd, Saber N. Elaydi, and G. Ladas (eds.), *Proceedings of the Sixth International Conference on Difference Equations*. Augsburg, Germany: CRC Press.

Index

For the benefit of digital users, indexed terms that span two pages (e.g., 52–53) may, on occasion, appear on only one of those pages.

206   INDEX